TECHNICAL REPORT

Outcome Evaluation of U.S. Department of State Support for the Global Methane Initiative

Nicholas Burger • Noreen Clancy • Yashodhara Rana • Rena Rudavsky
Aimee E. Curtright • Francisco Perez-Arce • Joanne K. Yoong

Sponsored by the U.S. Department of State

Environment, Energy, and Economic Development Program

This research was sponsored by the U.S. Department of State and was conducted in the Environment, Energy, and Economic Development Program within RAND Justice, Infrastructure, and Environment, a division of the RAND Corporation.

Library of Congress Cataloging-in-Publication Data

Burger, Nicholas.
  Outcome evaluation of U.S. Department of State support for the Global Methane Initiative / Nicholas Burger,
Noreen Clancy, Yashodhara Rana, Rena Rudavsky, Aimee E. Curtright, Francisco Perez-Arce, Joanne K. Yoong.
    pages cm
  Includes bibliographical references.
  ISBN 978-0-8330-7672-4 (pbk. : alk. paper)
  1. Greenhouse gas mitigation—Government policy—United States. 2. Greenhouse gas mitigation—United
States—Evaluation. 3. Methane—Environmental aspects. 4. United States. Dept. of State—Evaluation. I. Title.

  TD885.5.G73B87 2013
  363.738'7460973—dc23

                                                                                            2013000708

The RAND Corporation is a nonprofit institution that helps improve policy and decisionmaking through research and analysis. RAND's publications do not necessarily reflect the opinions of its research clients and sponsors.

RAND® is a registered trademark.

Published 2013 by the RAND Corporation
1776 Main Street, P.O. Box 2138, Santa Monica, CA 90407-2138
1200 South Hayes Street, Arlington, VA 22202-5050
4570 Fifth Avenue, Suite 600, Pittsburgh, PA 15213-2665
RAND URL: http://www.rand.org/
To order RAND documents or to obtain additional information, contact
Distribution Services: Telephone: (310) 451-7002;
Fax: (310) 451-6915; Email: order@rand.org

Preface

Methane emissions account for approximately one-third of anthropogenic climate forcing, or the heat-trapping effect of greenhouse gas emissions. Reducing methane emissions has increasingly become a goal for governments concerned about climate change. Because of the value of methane as a fuel (it is the main component of natural gas), consumers and producers have been interested in both the economic value and the environmental benefits of reducing methane emissions. This report evaluates U.S. Department of State (DoS) support for the Global Methane Initiative (GMI). GMI is an international partnership program that promotes cost-effective methane recovery and use by supporting public- and private-sector emissions reduction efforts. DoS—specifically, its Bureau of Oceans and International Environmental and Scientific Affairs (OES) and Office of Global Change (EGC)—has supplied funding to GMI totaling $27 million between fiscal years 2006 and 2010 and requested an evaluation of the activities and outcomes that it supported in whole or in part during that period.

The evaluation used quantitative and qualitative methods to describe and assess the effect of DoS support for GMI's methane reduction efforts. We also provide recommendations for the program and future evaluation efforts. Although the primary audience for this report is OES/EGC staff, the results should also be of interest to policymakers and stakeholders who are interested in voluntary actions by the public and private sectors to reduce greenhouse gas emissions.

The RAND Environment, Energy, and Economic Development Program

This research was conducted in the Environment, Energy, and Economic Development Program (EEED) within RAND Justice, Infrastructure, and Environment (JIE). The mission of RAND JIE is to improve the development, operation, use, and protection of society's essential physical assets and natural resources and to enhance the related social assets of safety and security of individuals in transit and in their workplaces and communities. The EEED research portfolio addresses environmental quality and regulation, energy resources and systems, water resources and systems, climate, natural hazards and disasters, and economic development—both domestically and internationally. EEED research is conducted for government, foundations, and the private sector.

Questions or comments about this report should be sent to the project leaders, Nicholas Burger (Nicholas_Burger@rand.org) or Noreen Clancy (Noreen_Clancy@rand.org). Information about the Environment, Energy, and Economic Development Program is available online (http://www.rand.org/jie/research/environment-energy.html). Inquiries about EEED projects should be sent to the following address:

Keith Crane, Director
Environment, Energy, and Economic Development Program, JIE
RAND Corporation
1200 South Hayes Street
Arlington, VA 22202-5050
703-413-1100, x5520
Keith_Crane@rand.org

Contents

Figures and Tables

Figures

Tables

Summary

Methane is a greenhouse gas (GHG) that has more than 20 times the warming power of carbon dioxide (CO_2) but remains in the atmosphere for a shorter amount of time.[1] Methane emissions are released during the course of a wide range of activities: the production and transport of coal, natural gas, and oil; raising livestock and other agricultural practices; and the decay of organic waste in municipal solid waste landfills and some wastewater treatment systems. In 2004, 14 countries came together under the leadership of the United States to launch the Methane to Markets Partnership. The program was relaunched in 2010 as the Global Methane Initiative (GMI). GMI promotes cost-effective, near-term methane recovery internationally through partnerships between developed and developing countries, with participation from private sector, development banks, and other governmental and nongovernmental organizations. GMI is part of the U.S. strategy to address GHG emissions and their impact on climate change.

As one of the two primary U.S. agencies participating in GMI, the U.S. Department of State (DoS)—specifically, its Bureau of Oceans and International Environmental and Scientific Affairs (OES) and Office of Global Change (EGC)—requested a study to "document and evaluate programmatic activities and outcomes relative to the contributions of OES/ EGC funding from fiscal years 2006 through 2010." OES/EGC requested an evaluation that described the value added of DoS contributions to the program, including a discussion of the countries and programmatic themes that were supported as a result of OES/EGC funding. They also requested that the evaluation apply a mixed-methods approach, using both quantitative and qualitative information, to document and illustrate program outcomes, including information from in-country site visits. DoS commissioned the RAND Corporation to conduct this assessment.

U.S. Government Support for GMI and the Role of DoS

GMI is a voluntary program that facilitates partnerships between member countries and private organizations, and the U.S. government (USG) provides financial and technical assistance to support the program and its goals. GMI's aims are to reduce methane emissions by raising global awareness about methane challenges and solutions, reducing institutional barriers, promoting learning, and facilitating knowledge-sharing. GMI is focused on reducing meth-

---

[1] Based on its 100-year global warming potential.

ane emissions across four sectors: agriculture, coal mines, landfills, and oil and gas systems.[2] USG funding supports activities across these four sectors, including feasibility studies, training workshops, demonstration projects, conferences, knowledge-sharing and dissemination opportunities, and efforts to facilitate technology transfer.

USG-supported activities promote methane reduction both directly and indirectly. For example, methane recovery demonstration projects reduce emissions directly, and those reductions can be measured. GMI works to facilitate emissions reductions indirectly, too. For example, some communities or organizations are unaware of the potential impact of methane reduction projects or do not know how to obtain the necessary financial and government assistance to initiate such a project. USG-supported educational efforts often take the form of meetings, conferences, training sessions, and workshops. These types of activities help participants share knowledge, build technical capacity, and promote other indirect outcomes that contribute to reductions in methane emissions.

GMI is led by a steering committee and four technical subcommittees (one for each sector), which include representatives from GMI partner countries. Both DoS and the U.S. Environmental Protection Agency (EPA) sit on the GMI Steering Committee, of which EPA serves as the chair. The committee provides overall direction to GMI. As a U.S. representative on the Steering Committee, DoS works with the other members to ensure that efforts undertaken by GMI are the best way to advance the program's goals and objectives. DoS also brings the U.S. foreign policy perspective to bear when guiding GMI's programmatic activities and strategic direction. GMI's Administrative Support Group (ASG) is hosted by EPA and serves as the secretariat, the main organizing and coordinating body. DoS assists the ASG with diplomatic interactions with partner countries and in identifying and engaging new potential partner countries.

Evaluation Approach

We evaluated DoS contributions to GMI using a mixed-methods approach that combined quantitative and qualitative data to characterize the DoS resources provided to GMI in fiscal years (FYs) 2006–2010, to identify the activities that GMI conducted with DoS support, and to assess the resulting achievements. Our focus was on DoS valued added—the additional benefits of the department's financial and nonfinancial contributions above and beyond other USG and non-USG support.

To assess value added, DoS contributions must be examined in the context of the overall program, since GMI is an integrated effort of DoS, EPA, and other stakeholders. We attempted to capture DoS contributions to GMI in two ways. First, we considered its share of the total financial support provided by the USG. We argue that DoS ought to be credited with at least the share of outputs and outcomes proportionate to its financial contribution. Second, we identified specific or unique contributions that DoS has made to the program, such as foreign policy guidance or flexible travel support, which other USG funders have been less able to provide.

---

[2]   In 2011, GMI added a wastewater systems sector, but because wastewater activities were part of the landfills sector during the study period, in this report we restrict our focus to the four original sectors.

We first examined the GMI program as a whole. We reviewed the financial and technical resources that the USG has contributed to the program (the inputs), the activities that have been undertaken on behalf of the program (the outputs), and results of those activities (the outcomes). We focused specifically on DoS contributions, including funding and strategic guidance. Because it is difficult to measure some outcomes, especially indirect outcomes, we used both quantitative and qualitative data to assess the activities and outcomes tied to OES/EGC funding and the value added of that support. We analyzed the available quantitative information and supplemented it with qualitative information from interviews and site visits.

Figure S.1 shows a simplified diagram of the key evaluation features in the context of the basic GMI program structure. Our evaluation focused on GMI inputs, outputs, and outcomes (shown at the top of the figure), which are related to specific program components (shown in the second panel). We drew on both qualitative and quantitative data (the third panel in the figure) to assess funding and strategic support (inputs), activities (outputs), and emissions reductions, institutional changes, and policy effects (outcomes). The final panel shows the primary sources of data on which we drew to assess each program component. The bolded boxes and text indicate the main focus of our evaluation, which was to assess DoS contributions to GMI, although these contributions are an integral part of the overall program. The red text indicates the core evaluation metrics, which we describe next.

To assess outcomes, we focused on a set of five evaluation metrics (shown in red text in Figure S.1). We drew on two sources to define the core evaluation metrics. First, we consid-

Figure S.1
Evaluation Framework

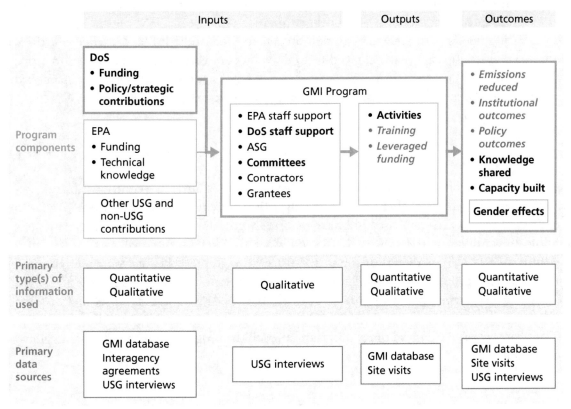

RAND *TR1250-S.1*

ered GMI's contributions to the four OES/EGC performance indicators, as outlined in the evaluation solicitation: policy outcomes, training, institutional capacity, and emissions reductions. We added to that set the metrics that the USG identifies as central to assessing its GMI support: emissions reductions, leveraged funding, and training. Because there was overlap between these sets of metrics, we consolidated them into five categories:

1. emissions reductions
2. leveraged funding
3. training
4. institutional outcomes
5. policy outcomes.

We also assessed the gender dimensions of these outcomes, where applicable and feasible, as requested in the solicitation. We summarized the metrics emissions reductions, leveraged funding, and training using EPA data. We were able to assess policy outcomes to a limited extent based on data and information gathered during site visits in three countries and from our interviews. Information to evaluate institutional outcomes, however, was almost nonexistent.

We organized the evaluation around the quantitative and qualitative data we compiled and collected, which provided complementary yet distinct insights into DoS contributions to GMI. For quantitative data, we compiled information on the amount of OES/EGC funding that was provided to GMI compared to total USG expenditures on the program for FYs 2006–2010. We also pulled data from EPA's GMI database—a system for tracking GMI activities and associated funding—on all GMI activities and outcomes funded by DoS (as part of USG contributions to GMI), by year, sector, and country.[3] For qualitative data, we examined program documentation, reviewed programmatic guidance in DoS funding documents, conducted interviews with relevant program officers in DoS and EPA, and carried out three country site visits. This approach allowed us to draw on complementary data sources to assess DoS contributions in terms of the aggregate share of GMI activities funded, their associated outcomes, and the specific administrative and programmatic contributions from DoS.

Findings

During the course of this evaluation, we found evidence that GMI has contributed to reducing emissions of methane. Of the approximately 2,000 activities initiated between FYs 2006 and 2010 in EPA's GMI database, 542 have reportedly contributed to reductions in emissions totaling 203 million metric tons of CO_2 equivalent (MMTCO$_2$e), equal to approximately one-third of total U.S. methane emissions in 2010. Although many factors contribute to emissions reductions, the scale of the decline in emissions reported by GMI is large. The actions of the international community also provide evidence that GMI is viewed as a useful effort. The number of countries that are members of GMI increased from 14 in 2004 to 41 in 2011, suggesting that there is substantial interest in the approach the program takes to addressing global climate change.

---

[3] EPA refers to this database internally as the Customer Relationship Management database, or CRM.

DoS has played a major role in providing USG support for GMI, especially in areas where DoS has stronger expertise than EPA, such as diplomacy and foreign policy guidance. DoS financial contributions to GMI have been substantial, accounting for slightly more than half of all USG funding for the program. The activities supported by DoS contributions—along with EPA's appropriations and other USG financial support—have contributed to approximately 150 MMTCO$_2$e in methane emissions reductions of the 203 MMTCO$_2$e reported by GMI, although we acknowledge that some of these data are difficult to verify.

Funding provided by DoS in conjunction with funding and technical assistance from EPA supported as many as 2,000 GMI-related activities, from training to reduce methane emissions from municipal waste facilities to pilot projects to reduce the leakage of natural gas from pipelines and coal mines. Without the funding—and the greater flexibility that DoS has than EPA to make some types of expenditures for program support—GMI's ability to pursue these activities and bring about the associated reductions in GHG emissions would have been greatly curtailed.

Quantitative Findings

Between FY 2006 and FY 2010, total USG funding for GMI was approximately $54 million; EPA and DoS were the primary funders. DoS monetary contributions were approximately $27 million, or 52 percent of total GMI funding. USG funds were used to support GMI activities around the globe, ranging from large-scale conferences and expositions to small-scale projects to reduce agricultural methane in developing countries. These activities resulted in both direct outcomes (e.g., reduced emissions) and indirect outcomes (e.g., improved technical capacity), although we can quantify the latter only through levels of output, such as people trained. EPA's GMI database captures approximately 2,000 GMI-related activities between 2005 and 2011 and directly associates USG funding with 1,095 of those activities.[4] According to EPA, many of the remaining 900 activities were also supported by DoS and EPA funding, but the funding information is missing or incomplete. Activities may have also been funded by GMI partners, Project Network members, or other stakeholders.

The 2,003 activities undertaken by GMI between 2006 and 2011 were relatively consistent in terms of sector served, project type, and region. GMI funding from all donors is co-mingled; thus, the GMI database did not allow us to ascribe specific funding amounts to specific activities. Consequently, we ascribed DoS value added to GMI activities and associated quantitative outcomes (emissions reduced) based on the 52 percent of total funding that DoS provided in support of GMI.

According to the GMI database, between 2006 and 2011 more than 15,000 people participated in GMI activities. We have categorized this participation by type of activities to focus on those that were designed specifically to promote learning or capacity-building. This subset of activities included approximately 45 percent of all recorded participants, with approximately 6,900 people reported to have attended a GMI-affiliated training session or workshop during our study period. Nearly all of these training sessions or workshops were funded at least in part by DoS and EPA. Of the 203 MMTCO$_2$e in methane emissions reductions recorded by GMI, the 532 USG-supported activities accounted for 146 MMTCO$_2$e, equal to about a quarter of all methane emissions from the United States in 2010.

---

[4]   In our study, we included data for 2011 because DoS FY 2010 funding supported activities in calendar year 2011.

Some GMI activities do not directly lead to emissions reductions: They are designed to build local capacity and transfer knowledge, which can lead to the spread of technologies and changes in business and other practices that could result in emissions reductions. If we restrict the number of activities we credit for reducing emissions to only those that were designed to lead directly to emissions reductions, then 540 (of 1,271, or 42 percent) of these activities directly contributed to reduced emissions.

Qualitative Findings
Site Visits

The RAND team's site visits captured more detailed qualitative evidence of DoS contributions to GMI, albeit for a limited set of countries and activities. During our site visits to India, Mexico, and the Philippines, we interviewed 32 individuals involved with approximately 30 distinct activities and visited six field sites. The goal of the country site visits was to understand how GMI activities were executed locally, to assess the nature of relationships among several stakeholders, and to gather richer information than that available from the GMI database about the activities conducted and their effects. The site visits were particularly useful in providing insights into the outcomes of GMI-funded activities that did not directly result in emissions reductions. The site visits also provided an opportunity to collect data from respondents to validate recorded program data.

Most respondents working in the four GMI sectors reported that USG-funded activities in support of GMI had helped educate industry and government leaders about the potential benefits of reducing methane emissions and about the potential to use methane collected from these activities as a fuel. In India, respondents working in both the landfill and coal sectors mentioned that GMI had resulted in increased initiatives to reduce methane emissions. In the Philippines, capacity-building activities, such as training, helped institutionalize methane-reducing practices, and the country's involvement with GMI has encouraged several agencies to establish a national equivalent, the Philippine Methane Initiative, whose aim is to develop a nationwide strategic plan for methane recovery and capture across the Philippines.

Stakeholders from both the private and public sectors stated that they benefited from capacity-building activities, such as study tours, conferences, and workshops, which gave them opportunities to learn about new technologies and network with leading international participants in their fields. These outcomes are hard to quantify and thus missing in the GMI database, but they were frequently cited by respondents during our site visits. These programs provided respondents with exposure to ideas and approaches that they otherwise would not have encountered.

The majority of our interviewees reported finding cross-sector activities beneficial, and they hoped that there would be an increase in such activities in the future. Comparing results across the three country case studies, we found that sectors that are fragmented and involve several organizations (e.g., agriculture) have a greater need for networking support than other, more consolidated sectors (e.g., oil and national gas, often run by one national agency). Fragmented sectors may benefit from additional funding or a local GMI representative who is able to coordinate activities among private, public, and community partners—a strength we observed in the Philippines. Participants also felt that industries characterized by small private firms would benefit from more demonstration projects.

Programmatic and Strategic Support

In addition to its monetary contributions, DoS provides programmatic and strategic guidance to GMI. DoS was heavily involved in the creation of GMI and continues to play an ongoing role through its participation in the Steering Committee and ASG. During GMI's formation, DoS contributed its skills in drafting multilateral agreements, especially when crafting the chartering documents, and engaged its core competence in diplomacy and building consent to create the partnership. DoS also engages in outreach to non-partner countries that it feels would be effective additions to GMI. DoS is able to use its knowledge and expertise to identify the appropriate ministers and to pursue the approaches that are likely to appeal to specific countries. DoS is also able to apply its diplomatic skills and international relations expertise when providing strategic and programmatic guidance to GMI.

Based on statements from DoS and EPA staff, without DoS financial and strategic support, GMI would be a very different and much smaller program than it is today. In our assessment, without DoS support, GMI's scope would probably have been reduced by more than half, because there are fixed costs associated with administering the program. In addition, DoS provides strategic and foreign policy guidance that falls outside EPA's technical expertise, a unique contribution given GMI's international structure. This contribution was mentioned by both EPA and DoS staff.

Recommendations

Based on our understanding of how GMI operates, the evolving role of DoS in GMI, and the data collection and management systems EPA maintains to track, assess, and report program accomplishments, we drew up some recommendations for ways to enhance DoS contributions and value added to GMI. We also identified opportunities to improve GMI data collection, especially to support future program evaluation.

Soliciting Feedback from Project Participants

In our interviews, we found that local stakeholders were aware of problems in implementing projects but felt that they lacked avenues through which to convey these observations to the USG. DoS should consider supporting a process to expand the channels through which stakeholders can provide information to program leaders that will help improve GMI.

Assessing the Evaluation Metrics

Metrics such as *emissions reduced* are relatively easy to measure and closely align with GMI's goals. Other metrics, such as *capacity built*, also closely align with GMI's goals but are more difficult to measure. In contrast, some of the metrics concerning impacts related to gender do not necessarily align well, based on our conversations with a range of GMI stakeholders. Because measurement often drives program focus, relatively weak alignment between metrics and program objectives can potentially distort performance. Overly narrow metrics, with corresponding annual targets, may result in funding being driven toward projects that "count," such as training programs for women, rather than on efforts focused on education, knowledge transfer, or partnership-building, which may have a greater effect on the long-term goal of reducing methane emissions. Overly broad metrics may reward "quantity" rather than "quality." EPA and DoS should consider developing appropriately tailored evaluation metrics as the program moves forward.

Leveraged Funding

Leveraged funding is an important potential benefit of DoS support for GMI. DoS funding can encourage other public- and private-sector entities to contribute additional funding to efforts to reduce methane emissions. But leveraged funding is challenging to measure. Current information on leveraged funding in the GMI database appears to conflate funding that "leverages" U.S. funds ("augments or builds upon an activity or effort funded by the U.S.," as stated in EPA's leveraged funding methodology) and funding from any source other than the USG that supports methane reductions (EPA, 2011a). We recommend that EPA and DoS develop standards for how leveraged funding is identified, what constitutes leveraged funding, and how much of the funding is designated as leveraged. This would lend additional credibility to reports on leveraged funding.

DoS Should Maintain a Supporting Role

DoS has provided substantial funding to support GMI, and it has also made important strategic contributions. However, DoS has put relatively few restrictions on how its funding should be allocated (e.g., across sectors or regions), and it allows EPA to play a lead role in managing USG support for GMI. We view this flexibility as beneficial. We recommend that DoS continue to provide high-level guidance and support while allowing EPA to drive the process of identifying technical opportunities and guiding USG funding allocations to the maximum extent feasible.

Opportunities for Future Program Evaluation

Based on our assessment of DoS support for GMI, we identified three activities that could supplement a long-term evaluation strategy and provide near-term insights into GMI's effectiveness, potentially at a relatively low cost:

- Conduct targeted surveys of individuals participating in GMI activities to assess
  - the types of benefits that GMI stakeholders perceive to be most valuable
  - the types and extent of information that stakeholders gain through participating in GMI activities.
- Assess both successful and unsuccessful grant applications—those just above and just below the cutoff line. This exercise could identify the topics for which USG support is pivotal versus topics for which there are potentially other available funding sources, allowing EPA and DoS to better target their resources.
- Construct a logic model, a necessary step to facilitate a process evaluation of GMI, that examines whether the program's activities and outputs are in line with its mission and are helping GMI achieve its goals.

Acknowledgments

We thank GMI program officers Barbara DeRosa-Joynt and Andrew Eil at the U.S. Department of State for sharing their insights with us. Griffin Thompson, also at the Department of State, was helpful in defining the research process, as was Nancy Ahson in providing historical documentation and managing the grant. We also thank the staff of the Environmental Protection Agency who have worked in support of GMI for explaining program details, providing program documentation, and being responsive to our many requests and questions. In particular, we thank Paul Gunning, Pamela Franklin, Henry Ferland, and Monica Shimamura for their support. We are grateful to all those whom we interviewed during our site visits who were gracious with their time and attention. We also acknowledge comments and suggestions from our two reviewers, Shanthi Nataraj of RAND and Tom Lyon of the University of Michigan.

Abbreviations

ASG	Global Methane Initiative Administrative Support Group
BAI	Philippine Bureau of Animal Industry
CBM	coal bed methane
CDM	Clean Development Mechanism
CMM	coal mine methane
CO_2	carbon dioxide
co-op	cooperative
DAP	Development Academy of the Philippines
DoS	U.S. Department of State
DOST	Philippines Department of Science and Technology
EEED	RAND Environment, Energy, and Economic Development Program
EGC	Office of Global Change, U.S. Department of State
EPA	U.S. Environmental Protection Agency
FY	fiscal year
GHG	greenhouse gas
GMI	Global Methane Initiative
IAA	interagency agreement
IQR	identification, quantification, and reduction
JIE	RAND Justice, Infrastructure, and Environment
LFG	landfill gas
MMSCM	million metric standard cubic meters
$MMTCO_2e$	million metric tons of carbon dioxide equivalent
$MTCO_2e$	Metric tons of carbon dioxide equivalent

NGO	nongovernmental organization
NGSI	Natural Gas STAR International
OES	Bureau of Oceans and International Environment and Scientific Affairs, U.S. Department of State
ONGC	Oil and Natural Gas Corporation
PEMEX	Petróleos Mexicanos
PMI	Philippine Methane Initiative
TOR	terms of reference
UN	United Nations
USAID	U.S. Agency for International Development
USG	U.S. government
USTDA	U.S. Trade and Development Agency
VAM	ventilation air methane

CHAPTER ONE

Introduction

The Global Methane Initiative (GMI) is a voluntary international partnership program that promotes cost-effective, near-term methane recovery and reuse activities in developing and transition economies with participation from the private sector, development banks, and other governmental and nongovernmental organizations. Methane is a short-lived greenhouse gas (GHG) that has more than 20 times the atmospheric warming effect of carbon dioxide (CO_2). Recognizing the important role of methane in global warming and its potential use as a clean energy source, 14 countries came together in 2004 under the leadership of the United States to launch the Methane to Markets Partnership, which was relaunched as GMI in 2010.

The U.S. government (USG) has been a GMI partner since the program's inception in 2004. The U.S. Department of State (DoS), specifically, its Bureau of Oceans and International Environmental and Scientific Affairs (OES) and Office of Global Change (EGC), has been a major contributor to GMI, along with the U.S. Environmental Protection Agency (EPA). OES/EGC requested a study to "document and evaluate programmatic activities and outcomes relative to the contributions of OES/EGC funding from fiscal years 2006 through 2010."[1] The agencies also requested an evaluation that describes their value added to the program, including a discussion of the countries and programmatic themes that were funded as a result of OES/EGC support to GMI. The evaluation applied a mixed methods approach, using both quantitative and qualitative information, to document and illustrate program outcomes, including information from three site visits. We did not, however, evaluate overall program effectiveness (e.g., relative to other GHG reduction programs), nor did we assess whether GMI's process and focus was the right one, given its mission.

To assess value added, DoS contributions must be examined in the context of the overall program, since GMI is an integrated effort of DoS, EPA, and other stakeholders. Implicitly, we assessed GMI against a hypothetical version of the program that did not include DoS financial or strategic support. In other words, how have DoS contributions shaped the program that GMI is today? We assessed DoS contributions to GMI—funding along with programmatic and strategic support—using quantitative and qualitative information that we applied to multiple output and outcome measures. The remainder of this chapter includes a brief discussion of methane as a GHG; a summary of the origins and structure of the GMI program, including a financial summary of USG support for the program; and a discussion of RAND's approach to the evaluation.

---

[1] Taken from the request for applications announcement, funding opportunity number OES-OCC-11-004 (DoS, 2011).

Methane as a Greenhouse Gas

Methane is a short-lived greenhouse gas: It remains in the atmosphere for approximately 12 years. Even though it is short-lived, it has more than 20 times the atmospheric warming effect of CO_2, which has an atmospheric lifetime of about 100 years. Methane is released during the production and transport of coal, natural gas, and oil; from raising livestock and other agricultural practices; and from the decay of organic waste in municipal solid waste landfills and some wastewater treatment systems. In 2003, researchers at the Massachusetts Institute of Technology reported that feasible reductions in emissions of methane and other non-CO_2 gases over the next half-century can make a contribution to slowing global warming that is as large as or larger than reductions in CO_2 emissions (Reilly, Jacoby, and Prynn, 2003). Because methane is a primary component (70–90 percent) of natural gas, efforts to reduce methane emissions can take advantage of current technologies that capture and reuse the gas as a fuel, potentially bringing about cost-effective reductions in emissions.

In December 2009, the United States proposed a GHG emissions reduction target in the range of 17 percent below 2005 levels by 2020 and approximately 83 percent below 2005 levels by 2050. It made this pledge at the Fifteenth Conference of the Parties of the United Nations (UN) Framework Convention on Climate Change as part of the Copenhagen Accord involving GHG mitigation contributions by developed and key developing countries (see DoS, 2010). The U.S. strategy for addressing GHG emissions and their impact on climate change involves engaging in multilateral initiatives and partnerships, such as GMI.[2]

The Global Methane Initiative

The Methane to Markets Partnership was launched in 2004 as a multinational effort to reduce methane emissions through cost-effective recovery and reuse. The founding partner countries were the United States and 13 other governments; since 2004 membership in what is now GMI has grown to 41 countries.[3] (For a complete list of partner countries, see Appendix A.) Methane to Markets operated through 2010 when it was relaunched as the Global Methane Initiative—in cooperation with the Asian Development Bank, the Inter-American Development Bank, and the European Commission—with an expanded mission to promote methane emissions reductions, in addition to the economical use of methane. Although the relaunch involved expanding the program's scope and international participation, the program's basic structure and approach were maintained. For simplicity, we refer to the program as *GMI* in this report, regardless of the period being discussed.

DoS and EPA are the primary USG agencies involved in GMI. Representatives of these agencies serve in leadership roles and provide funding to support on-the-ground methane reduction activities. DoS employs its core diplomatic and foreign policy skills to work with its

---

[2] Another recent initiative that involves GMI is the Climate and Clean Air Coalition, which was launched in February 2012 and focuses on reducing short-lived atmospheric pollutants (methane, black carbon, and hydrofluorocarbons). Coalition partners include the United States, Bangladesh, Canada, Ghana, Mexico, Sweden, and the UN Environment Programme. Work on the international level is taking place through the Global Methane Initiative, the Montreal Protocol, the Arctic Council, and the Global Alliance for Clean Cookstoves.

[3] We use the phrase *GMI countries* to include individual partner countries and the European Commission, which itself is a partner and participates in GMI in a capacity similar to a country.

international partners in GMI. EPA provides expertise and leadership in the technical areas, as well as administrative support.

GMI has the goal of reducing GHG emissions by advancing cost-effective, near-term methane recovery and use as a clean energy source in four sectors: (1) landfills, (2) coal mines, (3) agriculture, and (4) oil and gas systems.[4] Partner countries engage with the private sector to bring together the technical and market expertise, financing, and technology needed to develop methane capture and use projects around the world. GMI also has the goal of accelerating the deployment of technologies and practices that reduce methane emissions, thereby stimulating economic growth while reducing the effects of climate change and helping to improve local environmental conditions.

GMI takes a holistic approach to methane capture and use by bringing together the needed expertise and public and private authorities. For an individual project, the program often links private-sector project developers and technology providers with local and national governments and financing institutions. GMI activities also include efforts to reduce market barriers to development, such as institutional and informational challenges, by providing training and capacity-building assistance, technology demonstration projects, and other tools and resources to disseminate information and expertise. GMI partner countries have encouraged the participation of other entities that may have an interest in methane capture and use projects, such as the private sector, financial institutions, and nongovernmental organizations (NGOs). These entities participate as Project Network members. As of 2011, there were almost 1,100 Project Network members.

GMI's goals can lead to both direct and indirect outcomes. When a methane recovery project is implemented, it leads to emissions reductions, which can be measured—a directly measurable outcome. However, since GMI also seeks to facilitate information-sharing related to methane recovery and reduction processes, indirect outcomes are also important. The program recognizes that some communities or organizations are unaware of the potential to cost-effectively reduce emissions of methane and the opportunities for obtaining financial and government assistance in implementing such projects. Information is often provided through meetings, conferences, training sessions, and workshops. These types of outputs can spur outcomes, such as improved technical capacity, that may lead to the implementation of a methane recovery project and subsequent emissions reductions. However, as stand-alone activities, they do not yield directly measurable reductions in emissions. Consequently, outcomes from these activities are referred to as *indirect outcomes*.

The Structure of GMI
GMI's organizational structure is guided by a steering committee that provides overall leadership, by technical subcommittees (one for each of the four sectors) that provide technical expertise, and by the Administrative Support Group (ASG), which provides administrative and logistical support. The GMI Steering Committee currently has representatives from 23 partners and is chaired by the United States (EPA, specifically). The committee makes decisions about GMI's membership, organizational structure, and major initiatives. Its four technical subcommittees are chaired by one or more representatives from GMI partner coun-

---

[4] Municipal wastewater was originally part of the landfills sector, but as of October 2011, GMI elected to spin off municipal wastewater into a separate, fifth sector. Because we focused on the 2006–2010 period, we did not separately consider the wastewater sector in our analysis and thus refer to only *four* GMI sectors in this report.

tries. The ASG manages program operations, organizes and supports most GMI meetings, and provides technical support for the Steering Committee. The ASG, currently run by EPA, has created and manages a database of GMI-related methane reduction activities.[5]

The GMI partner countries play an important role in setting overall GMI priorities. The program provides a structure for countries to identify and pursue methane emissions reductions, and the technical subcommittees are responsible for developing action plans that identify the opportunities, barriers, and needs in each sector. Partner countries both inform and draw from these plans. In addition, partner countries are asked to develop country-specific methane action plans. These are strategic documents that, similar to the sector plans, identify barriers, opportunities, and—particularly in the case of developing countries—areas in which the country needs assistance in pursuing emissions reductions. In the action plans, developed countries are asked to identify ways in which they can provide assistance, either through technical support or capacity-building. The GMI all-partnership meetings, which are usually held annually, provide an opportunity for partner-country representatives and the technical subcommittees to meet, report on their activities, and exchange information related to sectoral and national planning. The following section discusses the specific role of the United States in each of these functional areas of GMI.

Role of the United States and DoS in GMI

As a founding member, the United States played a key role in the creation of GMI and continues to play a major role through its leadership and active participation in the Steering Committee, technical subcommittees, and the ASG. EPA was able to bring its technical expertise from parallel domestic voluntary programs to assist with the initial formation of GMI: AgSTAR, the Coalbed Methane Outreach Program, the Landfill Methane Outreach Program, and Natural Gas STAR International (NGSI). DoS contributed its skills in drafting multilateral agreements, especially when crafting the charter documents, and engaged its core competence in diplomacy and building consent to create the partnership. Both EPA and DoS were engaged in outreach to non-partner countries that they felt would be effective GMI partners. EPA often approached countries in which it had existing relationships through other initiatives. In many cases, EPA sought the help of DoS to identify the right government officials and offices and to pursue the right approach that might appeal to that particular country.

DoS played an important role in the launch of GMI, providing policy and foreign affairs support to the new multicountry agreement. Since then, DoS has worked to promote GMI within its own organization by educating embassy staff and country desk officers (with varying degrees of success as a result of high turnover, specifically for embassy staff, who are typically on two-year assignments).

Although EPA and DoS constitute the bulk of U.S involvement with GMI (including financial support), other government agencies contribute in-kind support in the form of technical expertise; some agencies historically provided coordinated financial support for GMI activities. Representatives from the U.S. Department of Agriculture, the U.S. Department of the Energy, and the U.S. Agency for International Development (USAID) are currently mem-

---

[5] For details on GMI's structure, roles, responsibilities, and members of the various committees and groups, see GMI, undated(a).

bers of the technical subcommittees. The U.S. Trade and Development Agency (USTDA) provided financial resources in the first few years of the program.

Steering Committee

Both EPA and DoS sit on the GMI Steering Committee (chaired by EPA), which provides overall strategic and programmatic guidance to the program. As members of the Steering Committee, both agencies contribute to the creation and drafting of a series of white papers, produced in advance and discussed at each Steering Committee meeting. These papers are the primary mechanism to guide the future direction of GMI. While both agencies provide programmatic input to ensure that the activities the Steering Committee promotes are the best way to advance the goals and objectives of GMI, individually, each agency provides different skills. EPA primarily contributes its technical expertise from similar efforts in the United States and its knowledge of international technical projects. DoS brings the U.S. foreign policy perspective to bear when commenting on GMI programmatic activities and strategic direction. DoS is also able to leverage its skills in engaging the international community in thinking about the future direction of the program.

Technical Subcommittees

Representatives from EPA, the U.S. Department of Agriculture, USAID, and the U.S. Department of Energy sit on the four technical subcommittees and bring their experiences from their domestic programs to the GMI process.[6] There is substantial knowledge-sharing between EPA's companion domestic programs (specifically, AgSTAR, the Coalbed Methane Outreach Program, the Landfill Methane Outreach program, and NGSI) and GMI. EPA leverages these domestic programs to provide the accumulated knowledge and technical capacity to inform GMI activities. While DoS does not currently sit on the technical subcommittees, it has been instrumental in providing guidance and support as to how such international partnership meetings should be conducted.

Each of the technical subcommittees develops an action plan that serves as a roadmap for future activities in that sector. Over the past few years, the subcommittees have developed country-specific action plans. The plans include existing and future opportunities for methane capture and use, descriptions of available technologies and best practices, identification of key barriers and issues for project development, identification of cooperative activities to increase methane recovery and use in the target sectors, and discussions of country-specific needs, opportunities, and barriers.

Administrative Support Group

The ASG, led by EPA, is the secretariat, the main organizing and coordinating body for GMI. It is responsible for managing GMI meetings and events, producing outreach materials, processing new members, maintaining the GMI website, facilitating communication between the committees and the Project Network members, and acting as an information clearinghouse. DoS has assisted the ASG by reaching out to partner countries through diplomatic channels. For example, in preparation for the all-partnership meeting in October 2011, the DoS program

---

[6] EPA is the only U.S. agency represented on all four technical subcommittees and is co-chair of the landfill and coal mine subcommittees.

officer for GMI sent cables to partner countries inviting their appropriate ministries to attend the meeting.

U.S. Funding Contributions to GMI

We have illustrated how EPA and DoS contribute to USG efforts to assist GMI, including providing strategic and programmatic guidance, oversight, and knowledge of technical issues and international negotiations. The USG also provides funding to support methane capture and use activities. EPA plays the role of the lead USG agency for GMI. Within EPA, GMI is managed by a team in the Office of Air and Radiation's Climate Change Division. In this capacity, EPA collects and coordinates all USG funding for GMI and distributes those resources in the form of funding for GMI activities to support international methane reduction activities. DoS is the source of approximately 52 percent of these funds.

In 2004, the USG pledged up to $53 million over five years to help develop GMI. In 2010, it renewed that commitment with "at least $50 million over the next five years" (GMI, 2011, p. 1). Since fiscal year (FY) 2006, annual USG funding for GMI has been approximately $10 million, with most of the funds coming from EPA (37 percent) and DoS (52 percent) (see Table 1.1). USAID and USTDA provided financial resources in the first few years of the program. However, since 2008, their support has primarily been "in kind" through the provision of technical expertise and assistance on individual projects or representation on technical subcommittees, in the case of USAID.

GMI has a congressionally appropriated line item in the budget for EPA. The EPA contribution cited in Table 1.1 takes into account total enacted appropriations for GMI (e.g., including full-time-equivalent staff, contracts, and grants).

DoS financial support for GMI began in FY 2006 through interagency agreements (IAAs) with EPA. EPA (in its capacity as the lead for USG support for GMI) and DoS negotiated the language in the IAAs, which direct how those funds will be used to support GMI. The parameters for the IAAs for the FY 2006–FY 2010 period were relatively constant. For example, the IAAs allocated approximately 10 percent of DoS funds to ASG support and 90 percent to support methane reduction activities. According to the IAAs, travel funding was typically limited to no more than 10–15 percent of the total DoS budget outlined in the IAA.

Table 1.1
Share of Total U.S. Government Support for GMI, by Agency, FYs 2006–2010 ($ millions)

Agency	Fiscal Year					Total
	2006	**2007**	**2008**	**2009**	**2010**	
EPA	2.00	4.40	4.40	4.50	4.60	19.90
DoS	5.80	6.00	5.66	5.00	5.30	27.76
USAID	3.07	0.85	0	0	0	3.92
USTDA	1.79	0.48	0	0	0	2.26
Total	12.66	11.73	10.06	9.50	9.90	53.84
DoS percentage of total U.S. funding						52

But the IAAs have also evolved over time, becoming more specific in terms of the types of activities to be funded and the countries in which USG-support activities will be implemented. The IAAs also identify emissions reduction targets by sector for the associated set of activities.

DoS and EPA work to fund activities that reflect the most recent guidance from the Steering Committee in terms of program priorities for GMI and that are consistent with the opportunities identified in the sector action plans developed by the technical subcommittees and the country-specific action plans developed by partner countries. The list of activities and the breakdown of funding in the IAAs are estimates for what the upcoming year of activities will entail for GMI. There is some flexibility should program plans and priorities shift or near-term opportunities arise. There are some limitations on the funds as well; for example, DoS funds cannot be used to support activities in China (as of 2011). The DoS funds can be used over more than one year, which provides more flexibility for a program that has to engage international partners that may face other constraints that affect program planning and timing. EPA is more restrictive in how much of its GMI funds can be used for travel.

EPA, in its role as the lead USG agency for GMI, manages the EPA-appropriated funds and the DoS funds from the IAAs and makes final decisions about activities to support GMI. EPA considers the priorities set forth by the Steering Committee, as well as the opportunities identified in the sector and country-specific action plans, in selecting activities. It also has to weigh certain constraints—such as the limited travel dollars available and the restriction that prevents DoS funds from being spent in China—as it identifies the set of activities to be funded in support of GMI. EPA administers the funds through two mechanisms: a grant solicitation process (there had been four rounds at the time of our study) and a contracting process. Table 1.2 provides a breakdown of DoS GMI expenditures through contracts and grants—the two primary funding vehicles used to support all activities—and travel.[7]

The study solicitation requested that DoS contributions to GMI be evaluated over the FY 2006–FY 2010 period (DoS, 2011). Since DoS funds can be used across more than one year, we have included FY 2011 in the evaluation of inputs, outputs, and outcomes because some FY 2010 dollars were used to fund FY 2011 activities. As shown in Tables 1.1 and 1.2, DoS has been the source of approximately 52 percent of total USG support for GMI over that period.

Table 1.2
Breakdown of DoS GMI Expenditures ($ millions)

Category	Fiscal Year				
	2006	2007	2008	2009	2010
Travel	0.20	0.40	0.16	0.20	0.15
Contracts	3.80	3.40	3.30	4.00	2.00
Grants	1.60	2.20	2.20	0.45	2.80
Indirect	0	0	0	0.35	0.35
Miscellaneous	0.20	0	0	0	0
Total	5.80	6.00	5.66	5.00	5.30

---

[7] Grants are typically referred to as *assistance agreements* in official documents; for clarity, we use *grants* throughout this report.

Evaluation Approach

According to the solicitation for this evaluation, the "primary purpose of the study" was to "document and evaluate programmatic activities and outcomes relative to the contributions of OES/EGC funding" in FYs 2006–2010 and to ascertain the value added to GMI outcomes as a result of that funding (DoS, 2011).

We evaluated DoS contributions to GMI using a mixed-methods approach that combined quantitative and qualitative data to characterize the resources provided to GMI over the FY 2006–FY 2010 period, to identify the activities that GMI conducted using DoS support, and to assess the resulting achievements. Our focus was on DoS value added—the additional benefits of DoS financial and nonfinancial contributions above and beyond other USG and non-USG support.

Figure 1.1 shows a simplified diagram of the key evaluation features in the context of the basic GMI program structure. Our evaluation focused on GMI inputs, outputs, and outcomes (shown at the top of the figure), which are related to specific program components (shown in the second panel). We drew on both qualitative and quantitative data (the third panel in the figure) to assess funding and strategic support (inputs), activities (outputs), and emissions reductions, institutional changes, and policy effects (outcomes). The final panel shows the primary sources of data on which we drew to assess each program component. The bolded boxes

Figure 1.1
Evaluation Framework

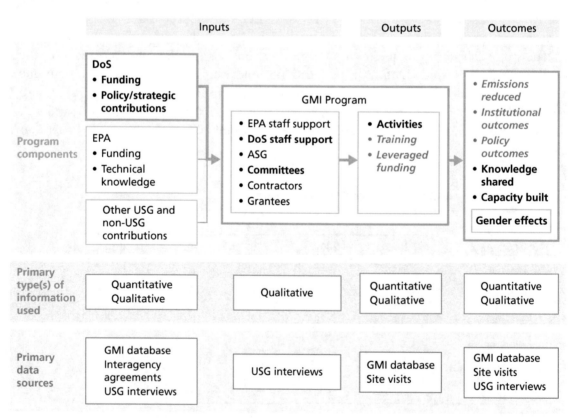

and text indicate the main focus of our evaluation, which was to assess DoS contributions to GMI, although these contributions are an integral part of the overall program. The red text indicates the core evaluation metrics, which we describe later in this chapter.

We first examined GMI as a whole. We reviewed the financial and technical resources that the USG has contributed to the program (the inputs), the activities that have been undertaken on behalf of the program (the outputs), and the results of those activities (the outcomes). We focused specifically on DoS contributions, including funding and strategic guidance. Because it is difficult to measure some outcomes, especially indirect outcomes, we used both quantitative and qualitative data to assess the activities and outcomes tied to OES/EGC funding and the value added of that support. We analyzed the available quantitative information and supplemented that with qualitative information from interviews and site visits to assess DoS contributions to GMI.

For the quantitative approach, we collected data on the amount of OES/EGC funding that was provided to GMI relative to the total GMI budget for FYs 2006–2010. We also collected information from GMI on the activities and outcomes that those funds supported by year, sector, and country.

Because GMI is a partnership meant to promote methane capture and use activities and reduce barriers to implementing these activities, rather than solely implement emissions reduction strategies, much of its effort and value was not captured in the quantitative data. Therefore, we also collected qualitative information about GMI, its activities, and outcomes. Collecting this qualitative information involved examining program documentation (e.g., GMI history, organizational structure, accomplishment reports, Steering Committee and subcommittee reports), programmatic guidance from the IAAs, interviews with relevant program officers in DoS and EPA, and three country site visits that included field visits to witness GMI-implemented activities and interviews with local funding recipients and other stakeholders.

Value Added

As part of the evaluation, we were asked to determine the "value added" to GMI from OES/EGC funding. According to the solicitation,

> OES/EGC funding for the GMI program has been in two key areas: (1) Administrative Support Group Activities including partnership meetings and outreach and communication; and (2) Project Development Activities in four target sectors including Agriculture, Landfills, Coal Mines, Oil and Natural Gas.

For the purpose of the evaluation, we defined *value added* as the effect of additional contributions to GMI by DoS above and beyond other USG contributions in terms of activities and outcomes. We acknowledge that this effect will not always be measurable, given the limitations of the existing data and the limited amount of new data collected for this study. Identifying value added is a descriptive and subjective process by nature, since systematic methods to discern value added (such as through comprehensive surveys of funding recipients and other stakeholders) were beyond the scope of this targeted evaluation.

DoS contributions must be examined in the context of the overall program, since GMI is an integrated effort among DoS, EPA, and other stakeholders. We therefore attempted to capture DoS contributions to GMI in two ways. First, we considered the department's share of the total financial support provided by the USG. We argue that DoS ought to be credited with at least the share of outputs and outcomes proportionate to its financial contribution. Second, we identified specific or unique contributions that DoS has made to the program, such as foreign policy guidance or flexible travel support, which other USG funders have been less able to provide.

We determined that the value added by DoS funding was approximately 52 percent of GMI's accomplishments, since DoS contributed 52 percent of total USG contributions to GMI (see Table 1.1). This constitutes one measurable dimension of DoS value added.

We made this assumption because the relationship between DoS and GMI has particular features that make it both difficult and, we argue, inappropriate to disaggregate program benefits or assess value added by activity. DoS provides funding to EPA (in its role as the lead USG agency) that is pooled with EPA congressional appropriations for GMI. DoS transfers come with few blanket restrictions—the only significant one being the restriction (as of 2011) that no DoS money be spent in or on China. Consequently, it is not possible to disaggregate the value added by DoS funding from the value added by EPA funding on an activity-by-activity basis.[8]

GMI Evaluation Metrics

To assess outcomes, we focused on five evaluation metrics (shown in red text in Figure 1). We drew on two sources to define the evaluation metrics for GMI outcomes. First, we considered GMI's contributions toward the four OES/EGC performance indicators, as outlined in the evaluation solicitation:

- amount of GHG emissions, measured in metric tons of CO_2 equivalent (MTCO$_2$e), reduced or sequestered as a result of USG assistance
- number of people receiving training in global climate change (by gender)
- number of laws, policies, agreements, or regulations addressing climate change proposed, adopted, or implemented as a result of USG assistance—specifically, those directly benefiting women or other marginalized groups
- number of institutions with an improved capacity to address climate change issues as a result of USG assistance—specifically, those serving women or other marginalized groups.

We also considered the metrics that the USG identifies for assessing its GMI support, which include emissions reductions (in tons of methane), leveraged funding (in U.S. dollars), and training (measured by attendance at GMI activities/events).[9]

---

[8]   In its GMI funding and activity tracking system, EPA is careful to track the money that DoS provides, the funding vehicles to which the money is assigned, and in what countries the money is spent. However, attribution breaks down at the level at which outcomes are recorded (i.e., the outcomes of specific activities). EPA does not break out DoS and EPA funding by activity. For these reasons, we believe that using the DoS share of total program funding is the most transparent and accurate quantitative approach to ascribing value added to DoS from its contributions to GMI. As noted earlier, we buttressed this quantitative approach with qualitative assessments based on interview data, site visits, and other sources of information.

[9]   These categories are from a document titled "US Government Efforts in Support of the Global Methane Initiative: Programmatic Metrics for Success," (EPA, 2011b), which is not available to the general public.

Given the overlap between the two sets of metrics, we consolidated them into the following five categories:

1. emissions reductions
2. leveraged funding
3. training
4. institutional outcomes
5. policy outcomes.

The first three metrics (emissions reductions, leveraged funding, and training) are all captured in EPA's GMI database. However, there are no systematic data—nor has GMI historically collected data—that would allow us to assess improved institutional capacity to address climate change, changes in the policy landscape, or gender impacts. Despite the lack of systematic data, we were able to assess policy outcomes to a limited extent based on data and information gathered during three country site visits and anecdotes mentioned in our interviews. In addition, the USG has reported qualitative outcomes in GMI's annual USG accomplishment reports (e.g., support for the Coal Mine/Coalbed Methane Clearinghouse in India).

Organization of This Report

Chapter Two discusses the quantitative results of our assessment, focusing on the project development activities that have been supported by GMI. These activities are captured in the GMI database and are categorized by type of activity, sector, country, and year. The data also include some outcome information, such as emissions reduction estimates. There are some limitations to these data, which are discussed in greater detail at the end of the chapter. Chapter Three provides insights from our qualitative data sources, with an emphasis on observations from the three country site visits. Chapter Four presents overall findings on the DoS value added to GMI. We also make recommendations for how DoS could improve the effectiveness of its contributions and for future evaluations, including suggestions for changes in the way in which programmatic data are collected to make GMI more amenable to evaluation in the future.

Quantitative Analysis of DoS Contributions to GMI Funding, Activities, and Outcomes

In this chapter, we summarize the existing *quantitative* program data, which allowed us to determine USG (and DoS) contributions to GMI, the range of activities undertaken with that support, and the resulting outcomes. The available quantitative data have limitations in terms of completeness. Chapter Three addresses gaps in the quantitative data with qualitative information from our interviews and country site visits. We begin this chapter by summarizing our evaluation approach using quantitative data, then we present summary information related to three of the evaluation metrics.

Quantitative Analysis Approach

In this section, we have three goals related to the evaluation framework presented in Figure 1.1: (1) describe in more detail USG funding for GMI, (2) summarize GMI activities, and (3) assess the evaluation metrics for which quantitative data can be used. The relevant evaluation metrics are *training, leveraged funding*, and *emissions reductions*. We do not consider institutional and policy outcomes in this chapter because there are no broad-based quantitative data available. Table 2.1 summarizes the evaluation metrics for which we had quantitative data and lists the specific ways we measured each.

The primary quantitative data source that we used to assess the evaluation metrics was the GMI database maintained by EPA in its role as the USG lead (e.g., tracking USG funding and activities) and in its capacity as head of the ASG to track certain activities for the entire initiative (e.g., contact lists, activities in the landfill and agricultural sectors). Throughout this section, we draw primarily on data from this database, which provides insights into GMI funding, activities, and outcomes. We used data on funding and activities to describe GMI's scope and characteristics related to DoS contributions, and we used other quantitative data to address the evaluation metrics outlined in Chapter One.

Table 2.1
Summary of Evaluation Metrics Addressed Using Quantitative Data

Evaluation Metric	How Measured
Training	Number of people attending training
Methane emissions reduced	Actual emissions reductions in metric tons of CO_2 equivalent ($MTCO_2e$)
Leveraged funding	Leveraged funding in dollars

As discussed in Chapter One, there are outcomes that are relevant to the success of GMI—outcomes that DoS and EPA staff emphasized as important features of GMI in our discussions—that we could not measure using existing quantitative data. These included capacity-building, knowledge transferred, national and subnational buy-in, and increased awareness of the importance of methane in addressing climate change. In addition, the DoS solicitation requested information on two metrics for which quantitative data were not available: the number of institutions with improved capacity to address climate change issues and the number of laws, policies, agreements, or regulations addressing climate change that had been proposed, adopted, or implemented. Where possible, we provide observations on these outcomes using qualitative data in Chapter Three. We also provide recommendations for ways to assess these outcomes more systematically in the future in Chapter Four.

We emphasize that data on activities completed (a measure of program output) should not necessarily be treated as less important than *outcomes*, such as reductions in emissions. Many outcomes that are important to GMI are difficult or costly to measure, such as capacity-building, knowledge-sharing, and awareness. Given GMI's goals—and, more specifically, the goals of DoS and EPA—*activities completed* is a relevant measure of programmatic achievement in that it reflects effort and resources that contribute to achieving outcomes. As we report in Chapter Three, we saw evidence during our site visits that GMI activities are valued even when they do not directly reduce emissions.

The GMI Database and How We Used It

EPA uses its online GMI database to record and track information about GMI funding and activities.[1] The database is designed to work as a clearinghouse for information about global methane reduction activities. It is the best centralized information source regarding GMI activities. Although the database was not specifically designed to be used as an evaluation support tool, it provides crucial information about the program's activities and accomplishments. Because GMI is a global program and staff work from non-U.S. locations, the online GMI database serves as a way by which EPA staff and contractors can easily input and retrieve information related to GMI activities.

A variety of methane reduction activities and projects are implemented under the GMI umbrella, and EPA tries to include all relevant activities in the database. This includes USG-funded efforts, but EPA also invites partners and Project Network members to submit information on methane reduction efforts that do not receive USG funding. Because GMI is voluntary in nature, all information is provided on a voluntary basis, which can lead to data gaps and potential variability in data quality and reliability. As EPA notes, this "makes robust data collection challenging as it relates to the activities and investments of Partners outside of the US" (EPA, 2011b).

The GMI database is a relatively new management system for EPA. Because the database is still being populated and refined, it has some limitations, which EPA is actively addressing. For example, older records in the database may lack information about the activity's start or end date. In completing our evaluation, EPA worked with us to compile new records and update existing records to allow us to use the data more effectively. This iterative process significantly improved the database's quality and completeness.

---

[1]   EPA refers to this database internally as the Customer Relationship Management database, or CRM.

One of EPA's main uses for the database is to track funding associated with GMI activities. Some of the most complete data in the GMI database relate to funding vehicles.[2] EPA records USG allocations to GMI activities through the "funding vehicle" variable in the database. Each funding vehicle reflects a discrete amount of EPA or DoS money (or both), which can fund several activities across various sectors, countries, and activity types. Because one funding vehicle can be used for several activities, it is impossible to use the data in the GMI database to determine in a straightforward way the amount spent by EPA or DoS on individual activities.[3] Thus, we used the funding vehicle data to determine GMI funding over time and across regions. Regional designations are based on the main country designated for each funding vehicle. We mapped countries to regions using the World Bank's regional designations. The region classifications are shown in Appendix A.

We then turned to the activity-level data as a unit of analysis to assess activities completed, training, and reductions in methane emissions. Activity-level data allowed us to categorize GMI activities along several dimensions, including sector, type, start year, and region. Sectors, activity types, and start years are all defined in the GMI database, but there are also several "cross-sector" designations. For clarity, we have collapsed them into a single "cross-sector" category. The activity data also provided information on training, event participation, and emissions reductions associated with each activity.

To capture as much information as possible about GMI and its activities, we downloaded several "flat" tables from the database and linked them together.[4] These tables covered funding, activities, and emissions. We downloaded the following six tables: (1) "Funding Vehicle," (2) "Funding Source," (3) "Funding Allocations," (4) "Funding Vehicle with Country," (5) "Activities," and (6) "Emissions." We merged the "Activities" table with the "Funding Vehicle" table through the "Activity Primary Funding Vehicle" variable.[5] We also merged the "Emissions" table and the "Activities" table through the "Activity Name" variable.[6] Although we pulled data from the GMI database repeatedly throughout the evaluation, we report numbers from our final data pull, taken on March 19, 2012. In total, the GMI database contains information on approximately $41 million in GMI-related expenditures between FY 2006 and FY 2010 and approximately 2,000 activities between 2006 and 2011.

---

[2] The funding vehicle information (e.g., sources, amounts) was complete, but, as we discuss later in this section, these data were not always linked to activity data.

[3] Funding vehicle dollars sum to dollars provided to GMI through EPA and DoS. When we compared these amounts to the amount flowing directly to activities in the GMI database, there was a $9 million difference. This was either the result of missing "Primary Funding Vehicle" data in the "Activities" table or because some activities were not associated with a particular funding vehicle. As a result, we summarize funding from the database's "Funding Vehicle" tables rather than imputations from the "Activities" tables.

[4] We joined tables to ensure that we were not inadvertently missing entries as a result of the the vagaries of the "Salesforce" merge functions. Salesforce was the creator of the database.

[5] Our analysis presumes that each activity was funded by no more than one funding vehicle. Using this many-to-one match assumption, we linked "Activity Primary Funding Vehicle" to "Funding Vehicle" to follow flows of money.

[6] As discussed later in this report, this method of merging revealed funding vehicles without associated activities and activities without associated reductions in emissions. These gaps were not revealed using the online software's merge functions, suggesting that caution is needed when using the software to validate whether the database is comprehensive. Future evaluations should seek a comprehensive understanding of the database structure, particularly because the database continues to be developed and populated.

USG Financial Support

In Chapter One, we summarized USG funding for GMI based on aggregate administrative data (i.e., EPA documents and DoS IAAs). Here, we use GMI database data on *spending* to provide additional details on how funding was allocated over time, across countries and regions, and by sector. Table 2.2 summarizes GMI funding from the USG (EPA and DoS) for FY 2006 though FY 2010.[7] As shown in the final column, total GMI funding captured by the table is approximately $41 million, which is less than the total listed in Table 1.1. The discrepancy is because some funding vehicles—mechanisms used to apportion funding to sets of tasks and activities, such as contracts or grants—in the GMI database are missing fiscal year information and because the database does not quantify USAID and USTDA funding. In addition, the annual and agency-specific amounts do not match those in Table 1.1 because Table 2.2 captures allocations to funding vehicles rather than agency allocations to GMI.

The table further breaks out GMI funding by source (either EPA or DoS), across years, and shows that the relative share of funding from DoS has varied over time, from 44 percent (FY 2008) to 61 percent (FY 2007).

The last two rows in the table report the share of total funding allocated to task orders and grants. EPA allocates USG funding in support of GMI through both contracting and grant-making. Total USG funds allocated to grants varied over time, falling in years when EPA did not issue a call for proposals for grant funding. Later, we describe the difference between the two mechanisms and the relative importance of each.

In Table 2.3, we break out funding vehicle data by region across fiscal years, showing the percentage of annual funding allocated to each region. The table shows that most regions received relatively consistent funding between FY 2006 and FY 2010, with the largest share allocated to the Latin America and Caribbean region in all but one year (not including all-partnership funding). The share of funding dedicated to activities in the East Asian and Pacific

Table 2.2
GMI Funding, by Fiscal Year ($ millions)

Category	2006	2007	2008	2009	2010	Total
Total GMI funding	3.08	4.98	15.40	9.55	8.02	41.02
EPA funding	1.57	1.93	8.67	4.11	3.82	20.10
DoS funding	1.51	3.04	6.73	5.44	4.20	20.93
DoS share of total (%)	49	61	44	57	52	51
Contracts share of total (%)	100	40	65	66	74	66
Grants share of total (%)	—	48	33	33	14	31

SOURCE: "Funding Vehicle" table, GMI database, as of March 19, 2012.
NOTE: The table excludes funding amounts with missing fiscal years. The share of contracts and share of grants categories do not add to 100 percent because a small portion of funding goes to other funding mechanisms. The GMI database also does not include funding from USAID and USTDA; thus, funding from these agencies is not included in the table.

---

[7] Fiscal years were calculated based on the funding vehicle start date.

Table 2.3
GMI Funding, by Region and Fiscal Year (% of annual total)

Region	Fiscal Year				
	2006	**2007**	**2008**	**2009**	**2010**
Partnership-wide	26	22	27	25	34
East Asia and Pacific	6	17	33	28	11
Latin America and Caribbean	38	28	14	29	27
Europe and Central Asia	16	22	12	13	23
South Asia	14	9	13	5	4
Sub-Saharan Africa	0	2	1	1	1
Middle East and North Africa	0	0	0	0	0
North America	0	0	0	0	0
Total	100	100	100	100	100

region has increased since the early years of GMI, with the exception of FY 2010. In contrast, funding for South Asia fell. The Middle East and North Africa region received little GMI funding (a total of $62,000 between FY 2006 and FY 2011). North America also received little GMI funding because Mexico—which has received substantial GMI funding—is part of the Latin America and Caribbean region. Sub-Saharan Africa also received little funding, in part because, as of 2011, there were only three Sub-Saharan partner countries in GMI; of these, only one (Nigeria) is a major methane-emitting country.

The Role of Contracts and Grants

EPA uses two primary mechanisms to allocate USG funding in support of GMI: contracts and grants.[8] Contracts are carried out in the form of task orders by EPA service contractors to perform specific activities that EPA (with input from DoS or GMI partners) has identified to support the GMI partnership and to achieve program outcomes. In contrast, the grant mechanism is used to support activities that are proposed and carried out by eligible organizations in GMI partner countries that share GMI's goals to reduce emissions of methane. EPA's grants are allocated through a competitive solicitation process. One difference between contracts and grants is that EPA has direct control over the activities completed through contracts. EPA has less substantive involvement in grants, although the grant agreement may specify the funding recipient's reporting and other obligations.[9] With contracts, the issuing agency can target a specific methane reduction activity in a specific country that the agency believes is needed, will help attract other investment, or will catalyze a sector. With grants, the agency

---

[8]   EPA includes grants in the more general category of *assistance agreements*, which can include grants or cooperative agreements. We use *grants* throughout, because that is the term used in the GMI database and in the funding documents we reviewed. EPA also uses other funding mechanisms in its work to support GMI, including IAAs (e.g., with DoS), which we discuss elsewhere, and miscellaneous obligation documents, which we do not discuss.

[9]   EPA has more substantive involvement in activities managed through cooperative agreements than it does for grants, but, in general, both mechanisms provide the funding recipient with greater opportunity to design and direct the activity than is the case with a contract.

chooses whether or not to fund a proposal that has been submitted. Other than the guidance and review criteria provided by the request for proposals (described later), EPA has less control over the types of proposals it receives and the specific activities proposed.

We provide more detail on the grant-making process because grants offer an important opportunity for organizations in GMI member countries to get funding to support their activities. Grants are not more important than contracts, but they represent a unique opportunity for EPA and the organizations it funds. Between FY 2006 and FY 2010, EPA issued four rounds of requests for proposals for grants to fund activities that help advance methane recovery and use through GMI. A wide range of activities were considered eligible for funding, such as feasibility studies, demonstration projects, projects that help foster bilateral or multilateral cooperation, measurement and estimation studies, studies that help improve estimation procedures, and awareness and capacity-building projects. All organizations except private for-profit and state-based for-profit organizations were eligible for funding.

In total, approximately $16 million was provided to fund 100 grants across 22 GMI member countries. DoS funding was used for 63 grants, with total funding of $9.1 million. EPA funds were used for 34 of the activities, with total funding of $5.7 million, and three grants ($0.9 million) were supported by joint DoS/EPA funding. Roughly equal numbers of grant activities were funded in all five years, except in 2010, when only two continuing activities received grant funding. With respect to both the total number of activities and the amount of funding received, China was the biggest beneficiary. In China, 23 grants were funded, for a total of $3.7 million.

Activities and Participation

EPA, along with its country partners and network members, carries out a wide range of activities under the auspices of GMI—from feasibility and measurement studies to training workshops and meetings. In this section, we present data on the activities undertaken under GMI (disaggregated along several dimensions). We also assess the number of people trained through GMI activities, one of our evaluation metrics.

Although a majority of activities captured in the GMI database were undertaken by EPA, its contractors, or grant recipients, there were some activities in the database that did not receive direct USG financial support. Furthermore, it was often not straightforward to determine which activities received USG support and which did not. Because of issues concerning the migration of data from an old database to the new one, the indicator for USG funding was incomplete.[10] According to our interviews with EPA staff, EPA considers all activities reported in the database as something GMI has "touched" in some way, even if the activities did not receive USG funding. Thus, GMI's association with the activity is the criterion for its inclusion in the database. With no way to cleanly differentiate USG-funded activities from non–USG-funded activities, combined with the inclusion criteria for the database, we took a broad view of what constitutes GMI-related activities. While we report aggregate summary information

---

[10] Of the 2,003 activities implemented between 2006 and 2011, 390 were missing information on whether the activity received USG funding. In addition, 1,095 activities were designated as USG-funded, while 518 activities were recorded as having received no USG funding.

for activities that were explicitly cataloged as receiving USG funding, most of the data we present covers all activities captured in the GMI database.

In the majority of cases, each activity recorded in the GMI database reflected a discrete activity that received support from EPA or DoS. One important exception was NGSI, which is considered part of USG-supported GMI efforts. All activities carried out under that program are collapsed into a single activity for the purposes of reporting. Consequently, the number of activities reported in the activities section slightly understated the actual number of individual activities carried out by GMI, as the number did not include the discrete activities undertaken under NGSI.

Activities

In this section, we summarize the activities recorded in the GMI database. We focus on trends in activities for two reasons. First, the activity data allowed us to conduct analyses that were infeasible with the funding data, such as more accurately accounting for sector-specific GMI support. Second, activities are the primary outputs for GMI; it is through these activities that GMI seeks to achieve such outcomes as reductions in methane emissions. Many of GMI's efforts are not designed to lead to emissions reductions directly; focusing only on outcomes— emissions—would not provide a complete picture of the accomplishments associated with DoS support.

Table 2.4 summarizes the GMI activities conducted between 2006 and 2011. The table shows that the GMI database includes just over 2,000 activities, an average of more than 300 activities per year. In the program's early years, however, the number of activities conducted was much lower. Moreover, in 2011, the database shows fewer activities conducted than either 2009 or 2010. As the second row indicates, the downward trend in 2011 is only partly due to fewer USG-funded activities. The overall decrease in activities may reflect delays in identifying and inputting non–USG-funded activities, since EPA must solicit input from country partners and network members to collect this information.

The distribution of GMI activities varied across regions and over time. Table 2.5 shows the share of USG-funded activities started each year in each geographic region. A large majority of activities were carried out in the East Asia and Pacific, Europe and Central Asia, and Latin America and Caribbean regions. Although annual shares of activities in these regions fluctuated over time, there are few notable trends. One exception is the Latin America and Caribbean region, which conducted a plurality of activities each year between 2006 and 2010. Consistent with the allocation of funding, GMI conducted few activities in the Middle East and North Africa, North America, and Sub-Saharan Africa regions.

Table 2.4
GMI Activities, by Year

Category	2006	2007	2008	2009	2010	2011	Total
GMI activities	199	177	255	460	626	286	2,003
USG-funded activities	91	109	114	313	276	192	1,095

SOURCE: "Emissions" table data joined with "Activities" table data, GMI database, as of March 19, 2012.

NOTE: The USG-funded category includes all activities designated as having received USG funding, but some activities that received USG funding may be excluded due to missing data. The table does not include 70 activities for which no start year was recorded. The data on USG support for activities was limited: for 469 activities, there was no indication of whether the activity received USG support.

Table 2.5
Share of GMI Activities, by Region and Year (%)

Region	2006	2007	2008	2009	2010	2011	2010 Methane Emissions Share
East Asia and Pacific	26	23	23	21	15	36	25
Europe and Central Asia	14	24	21	15	8	31	18
Latin America and Caribbean	46	43	40	50	71	24	14
Middle East and North Africa	0	0	0	0	0	0	8
North America	5	2	7	4	3	3	10
South Asia	5	6	4	9	2	4	12
Sub-Saharan Africa	4	2	4	1	0	1	12
Total	100	100	100	100	100	100	100

SOURCE: "Emissions" table data joined with "Activities" table data, GMI database, as of March 19, 2012. Information on the regional share of methane emissions is from EPA, 2006.

NOTE: The table reflects data on a total of 1,951 activities; 47 activities associated with non-country entities (e.g., the European Commission) have been excluded.

In the rightmost column in Table 2.5, we report each region's share of global methane emissions in 2010. Although activity allocations are not proportional to each region's share of global methane emissions, the regions that are the highest emitters (i.e., East Asia and Pacific, Europe and Central Asia, and Latin America and the Caribbean) conducted a large share of GMI activities. Sub-Saharan Africa, the Middle East and North Africa, and North America conducted a small share of activities relative to their share of global methane emissions. These disparities should not be overinterpreted, however, since each activity does not represent a comparable level of effort. Moreover, there are reasons beyond methane emissions levels that some regions receive more attention than others (such as predominant emitting sectors or available capacity). North America, for example, is a high-income, high-capacity region; the United States, the largest economy in this region, is served by EPA's domestic methane reduction programs. A large share of methane emissions in the Middle East and North Africa region comes from the oil and natural gas sector, which can be challenging to target because large gas companies require additional effort to bring on board to GMI.

We next break down activities by both sector and activity type. To show the overall distribution of activities across sectors and assess whether certain regions tend to host more activities in certain sectors, Table 2.6 presents the share of activities in each region that were allocated to each of the four GMI sectors. The table also includes ASG activities and cross-sector activities. In four of the seven regions, the landfill sector accounted for the majority of GMI activities. There were wide variations in the share of agriculture activities in the total number of activities, reflecting the important role agriculture plays in generating methane emissions in, for example, the East Asia and Pacific and the South Asia regions. In contrast, although a substantial share of methane emissions in Sub-Saharan Africa came from the agriculture sector (approximately 47 percent), only 16 percent of the region's activities were focused on agriculture.[11]

---

[11] Data on agriculture's share of total methane emissions in Africa are from EPA, 2006.

Table 2.6
USG-Funded GMI Activities, by Region and Sector (%)

Region	ASG	Cross-Sector	Agriculture	Coal Mines	Landfills	Oil and Gas	Total
East Asia and Pacific	0	2	37	26	26	8	100
Europe and Central Asia	0	7	30	4	47	11	100
Latin America and Caribbean	0	2	1	57	35	5	100
Middle East and North Africa	0	0	0	0	33	67	100
North America	1	26	31	14	13	14	100
South Asia	0	6	36	16	33	10	100
Sub-Saharan Africa	0	3	16	6	71	3	100
Total	0	36	17	4	35	7	100

SOURCE: "Emissions" table data joined with "Activities" table data, GMI database, as of March 19, 2012.
NOTE: The table reflects data on a total of 2,017 activities; 45 activities associated with non-country entities (e.g., the European Commission) have been excluded.

We next consider the types of activities implemented under GMI. As described earlier in this report, GMI supports different types of activities, some designed to reduce emissions directly and others geared toward building capacity to undertake future reductions. Table 2.7 shows trends over time and the overall distribution of types of activities. The share of activities for each type has been relatively stable over time, especially when we exclude activities related to the two large-scale GMI expositions. Overall, projects and studies constituted the largest share of activities, followed by meetings and posters.

Table 2.7
GMI Activities, by Type and Year (%)

Activity Type	2006	2007	2008	2009	2010	2011	Overall
Studies	11	40	25	17	24	26	23
Technical assistance/outreach	3	6	4	3	4	5	4
Workshops/training	6	7	7	8	6	14	8
Meetings	13	14	24	12	13	31	17
Expos	0	1	0	0	0	0	0
Expo posters	43	1	0	30	0	0	11
Other	1	2	2	1	0	1	1
Projects	24	28	38	27	51	21	35
Study tours	0	1	0	1	1	3	1
Total	100	100	100	100	100	100	100

SOURCE: "Emissions" table data joined with "Activities" table data, GMI database, as of March 19, 2012.
NOTE: Some activities that were large in scale but small in number (e.g., the 2010 expo in New Delhi) are not included in the table. The table reflects data on a total of 2,003 activities; 70 activities with no start years have been excluded.

Table 2.7 should be interpreted with caution, because the activities varied significantly in content and scope. For example, as noted earlier, the USG has supported two large-scale expositions since the inception of GMI, with a third planned for 2013 in Canada. The expos are large events that require significant coordination and financial support. They also represent substantial potential for learning, capacity-building, and raising awareness in a way that other meetings do not. On a more limited scale, the same is true of all-partnership meetings, such as the October 2011 meeting in Krakow, Poland. These major initiatives show up as a single activity in the database and, thus, in Table 2.7. In contrast, each of the dozens of expo posters—visual displays that summarize GMI-related activities to share knowledge or generate interest—created for the two GMI expos constitutes an individual activity. Summarizing activities can provide insight into GMI's evolving priorities, but data on the number of activities should not be interpreted as equivalent to the relative amount of effort or impact.

Between 2006 and 2011—through GMI—DoS and EPA supported or participated in approximately 2,000 activities across the four GMI sectors in all regions of the world, and DoS funding accounted for 52 percent of all USG financial contributions to support those activities.

Training

We measure training by the number of participants in GMI activities. The GMI database contains information on the number of people who have participated in GMI activities, either as counted directly by EPA or as reported by subcontractors, grantees, or GMI partner organizations. Two caveats are worth noting with regard to participation data. First, in many cases, EPA (or its partners) estimate attendance at GMI activities and, therefore, there will be some error in attendance numbers. We have no reason to believe that this error is anything but random; thus, across activities, we assume that the error averages to zero. Second, EPA counts total attendance at activities or events for which it has provided financial support or in which it was involved. For example, EPA provided funds for its staff to attend a major petroleum conference in 2011 and present on GMI; the database records 1,500 attendees for this event. Consequently, the *degree* of USG support or involvement across different activities can vary significantly.

According to the GMI database, between 2006 and 2011, more than 15,000 people participated in GMI activities. Some activities had recorded attendance as low as one person, while others, such as the 2011 Indonesian Petroleum Association's annual conference in Southeast Asia, had as many as 1,500 attendees. Table 2.8 summarizes attendance data for GMI and USG-funded activities. Although the number of attendees increased over the study period (exceeding 5,000 in 2011), the increase mostly reflects a rise in the number of activities. Table 2.8 shows that average attendance per activity has been relatively constant over time. Although total attendance appears to have increased, this apparent trend is at least in part the result of improved data collection in recent years, so the attendance numbers for training for earlier years are likely to be underestimates.

We break down participation by type of activity to focus on activities that were specifically designed to promote learning or capacity-building. The third row of Table 2.8 shows participation data specifically for activities classified as workshops or training sessions. This subset of activities includes approximately one-third of all recorded participants, with approximately 6,900 people reported to have attended a GMI-affiliated training session or workshop between 2006 and 2011.

Table 2.8
Total Participation in GMI Activities Recorded in the GMI Database

Category	2006	2007	2008	2009	2010	2011	Total
GMI activities	759	1,579	1,349	2,333	4,049	5,176	15,245
Average, per activity	69	44	30	33	45	53	43
Training/workshops only	146	485	746	1,489	2,300	1,516	6,682
USG-funded activities	209	1,515	1,289	2,298	4,013	4,461	13,785
Average, per activity	23	54	30	33	46	55	43
Training/workshops only	96	455	746	1,489	2,300	1,431	6,517

SOURCE: "Activities" table, GMI database, as of March 19, 2012.

NOTE: The database contained participation data for a total of 357 GMI activities, of which 109 were training or workshop activities. The database contained participation data on a total of 325 USG-funded activities, of which 104 were training or workshop activities.

As mentioned earlier in this section, although we report on participation in activities that were recorded as receiving USG funding, we emphasize that this information is incomplete. Nearly all participation data in the GMI database were associated with USG-funded activities. USG-funded activities saw approximately 14,000 participants over the seven-year study period. Participation in USG-funded activities followed the same increasing trend as overall participation. Nearly all participants in training sessions or workshops were attending USG-funded activities. Applying the DoS funding-based allocation rule, we would attribute approximately 7,300 participants to DoS funding across all activities, or approximately 3,500 participants in training activities or workshops.

Emissions Reductions

GMI's primary goal is to facilitate reductions in methane emissions. The program aims to achieve reductions in part by building capacity and facilitating knowledge-sharing, but USG funding also supports direct reductions in emissions through demonstration projects and feasibility studies.[12] In this section, we review the available outcome data and summarize GMI outcomes, including DoS contributions. We focus primarily on reductions in emissions, the principle GMI outcome, although there are important limitations to assessing GMI's contributions to reducing emissions. In addition, there are other relevant outcomes or suboutcomes, such as gender and policy effects, which we describe in more detail at the end of this section.

Measuring Reductions in Methane Emissions Using GMI Data

EPA collects information about GMI activities (e.g., via progress reports and other communications with grant recipients and contractors) to track reductions in emissions that have resulted from USG-supported activities. EPA, in its role as head of the ASG, also tracks emis-

---

[12] Although feasibility studies do not, in and of themselves, reduce emissions, we characterize them as "direct" because they focus on specific emissions reductions as opposed to building capacity or sharing broad-based knowledge.

sions reductions from GMI activities that did not receive USG support by requesting that GMI partners report their accomplishments to EPA.

Throughout this report, we report methane emissions in metric tons of carbon dioxide equivalent (MTCO$_2$e). Although methane has a shorter atmospheric lifespan than CO$_2$, it has a higher global warming potential, making it a more potent GHG. EPA follows guidance from the Intergovernmental Panel on Climate Change's *Second Assessment Report* (1996) and uses a 100-year global warming potential for methane of 21 (Houghton et al., 1996).[13] We follow EPA's assumption for ease of interpretation and comparison with previous GMI reporting, but we note that the *Fourth Assessment Report* (Forster et al., 2007) uses a revised 100-year global warming potential of 25, or approximately 19 percent higher than EPA's assumed value. In this respect, the emissions reductions that we report in terms of MTCO$_2$e are underestimates.

GMI tracks both *actual* and *potential* emissions reductions, where actual reductions are those that have been achieved to date and potential reductions are those that have been identified and could be achieved in the future if the projects are actually implemented. An example of potential reductions is a feasibility study that finds that a proposed landfill gas extraction project would reduce emissions by a specific amount, but the project has yet to be implemented. If the recommendations in the feasibility study were implemented, the landfill would achieve actual reductions that could be equal to, greater than, or less than the estimated potential reductions. We report actual emissions reductions achieved by activities during the study period (2006–2011) only. Because we focused on outcomes achieved to date, we do not include potential reductions.

Although we did not analyze potential reductions as part of this evaluation, because the focus was on emissions outcomes achieved, we do note that GMI and many of its activities involve efforts focused on identifying *opportunities* to reduce emissions. Studies that identify and estimate these reductions can act as catalysts for future projects that could lead to actual reductions. We learned in our interviews that estimates of potential reductions are often the way that GMI is able to engage both private and public sectors. These studies represent GHG-reducing opportunities and, often, cost savings, and they were cited by EPA program officers as a major factor in getting organizations to participate in GMI programs. In 2010, the GMI database recorded 29.5 million metric tons of CO$_2$ equivalent (MMTCO$_2$e) of potential reductions, a number consistent with totals identified in previous years. These estimates represent the future pipeline of opportunities to reduce methane emissions.

Another important feature of emissions reduction is the future stream of reductions from activities that have already been implemented. Some activities reduce emissions now and will continue to reduce emissions into the future. For example, a project to capture methane from a landfill will reduce emissions not only in the year the system is built, but also throughout the life of the landfill. We do not include future, *expected* reductions here, largely because we could not determine the lifetime of most activities. However, we emphasize that, for most activities, the actual reductions that we report can be viewed as annual reductions, which will lead to higher cumulative reductions over the lifetime of the project.

EPA has developed an approach to help ensure that reductions in emissions are recorded systematically. Each of the four GMI sectors draws on the comparable domestic sectoral program at EPA. Consequently, each sector has a separate (but related) approach for tracking and

---

[13] Over a 100-year period, one unit of methane has the same warming effect as 21 units of CO$_2$.

accounting for emissions reductions.[14] The coal mine, landfill, and oil and gas sectors have specific, written methodologies for "documenting and tracking methane emissions reductions." As of this writing, EPA was drafting a methodology document for the agricultural sector.

GMI collects and records emissions reduction data annually. In the case of the coal mine and landfill sectors, data collection and analyses make use of the GMI database, direct communication with partner countries, and data collected by EPA technical contractors. For the oil and gas sector, data are collected through two channels: annual reports by official partner companies and communications with non-official partner companies through less formal mechanisms (e.g., email, telephone, company presentations).

Documents describing the methodologies also outline approaches for "ascribing the relationship between these emissions reductions and the activities of U.S. agencies in support of the GMI." These numbers reflect the fraction of reductions for a specific activity that can be attributed to USG support; we refer to them as *attribution factors*. The attribution step is important. EPA recognizes that, in some cases, the USG was not the only source of support that contributed to an activity being undertaken and the associated emissions reductions realized. Where the USG played less of a role, the attribution factor is lower. If USG support was deemed the only reason that specific emissions reductions occurred, EPA allocated full credit to the USG—a 100-percent attribution factor. Table 2.9 summarizes the key attributes of the three sectors with emissions measurement documents. In the case of agricultural activities,

Table 2.9
Summary of Emissions Reduction Tracking Methodologies, by Sector

Sector	Qualifying Methane Sources	Attribution System	GHG Reduction Attribution Factors
Coal mines	Degasification systems from active underground coal mines Ventilation systems at active underground mines Abandoned (closed) underground mines Surface coal mines (pre-mine drainage)	Project- and site-specific Tiered	Three tiers, based on a percentage of the total annual emissions reduced: 90 percent 70 percent 40 percent
Oil and gas	Leaks Process venting System upsets (i.e., service disruption, maintenance activities)	Company-specific	All actual emissions receive 100-percent credit
Landfills	Land fill gas (LFG) flaring projects (direct methane reductions) LFG-to-energy projects (direct methane reductions) LFG-to-energy projects (indirect CO_2 reductions)	Project- and site-specific Tiered	Three tiers: 90 percent 70 percent 40 percent
Agriculture	—	—	100-percent credit

SOURCE: EPA emissions measurement methodology documents.

NOTE: The methodology document for the oil and gas sector refers to actual emissions only, not potential emissions. The entries for agriculture are blank because there was no published methodology document as of this writing.

---

[14] In this section, we summarize key components of the emissions calculation methodologies. These methodology documents are not disseminated publicly.

EPA's current practice is to ascribe 100 percent of all emissions reductions associated with GMI activities to USG support.[15]

Estimates of Reductions in Methane Emissions

Methane emissions reductions are quantifiable and comparable across settings and over time. However, it is not always feasible to calculate emissions reductions, either because of lack of resources or expertise or because an activity leads to emissions reductions only indirectly.[16] These difficulties can result in underestimates of total emissions reductions attributable to GMI, because some actual reductions are not captured in the GMI database. In contrast, these data could *overestimate* actual reductions if GMI attributes too large a share of those reductions to its effort through its attribution factor approach.

The GMI database includes approximately 2,000 activities initiated between 2006 and 2011, of which 542 (~27 percent) have recorded actual emissions reductions. Total reductions across all activities in the GMI database initiated between 2006 and 2011 were 203 MMTCO$_2$e, as summarized in Table 2.10. This is comparable to approximately one-third of total methane emissions in the United States in 2010. The number captured in the database includes activities reported to GMI for which the USG provided only some (or no) support. Using the attribution factor to identify only those activities that were associated with some USG support, total emissions reductions were 146 MMTCO$_2$e across 532 activities, as shown in Table 2.10.[17] "Attributable reductions" are calculated by multiplying the recorded emissions reduced for each activity by the attribution factor for that activity.[18]

DoS funding cannot be used to support activities in China, and the third and fourth rows in Table 2.10 summarize annual and total emissions reductions excluding activities that occurred in China. When we exclude the approximately 70 activities conducted in China, which accounted for 199 MMTCO$_2$e of total emissions and 86 MMTCO$_2$e of attributable emissions, we find that 474 non-China activities were associated with 84 MMTCO$_2$e of emissions reductions. Applying the attribution factor, there were 464 non-China activities that received USG support, associated with 58.8 MMTCO$_2$e of reduced emissions.

Only one-quarter of all activities had recorded actual emissions, but as Table 2.11 shows, this figure understates the share of activities that *could* lead to emissions reductions. Many types of activities, such as meetings or expositions, are not designed to reduce emissions directly, although these activities may help reduce emissions indirectly (e.g., by making knowl-

---

[15] As of this writing, EPA was drafting a methodology document for the agriculture sector that included tiered allocations.

[16] We use the term *indirect emissions* to refer to reductions that may result indirectly from an activity—such as a workshop, training program, or a meeting—for which it is not possible to calculate reductions in emissions; an organizer is unlikely to know whether someone who attended a workshop eventually installs a device to capture methane emissions. In contrast, GMI uses the term *indirect emissions* to describe reductions achieved by producing energy with captured methane and displacing other, potentially more carbon-intensive energy sources.

[17] According to conversations between the evaluation team and EPA staff, EPA believes that the existence of an attribution factor is the best indicator for whether an activity received USG support. This does not necessarily mean that the activity in question received USG *funding*, however. Because the USG funding indicator in the database was incomplete, we did not focus on emissions reductions for USG-funded activities, but we note that applying the USG funding indicator reduces total emissions to approximately 34 MMTCO$_2$e. The sharp reduction is because the GMI database reports only 24 activities that are designated as having received USG funding *and* for which there were recorded emissions reductions.

[18] Of the 532 activities that had an associated emissions factor, there were 455 activities with emissions factors of less than 100.

Table 2.10
Actual Emissions Reductions Recorded in the GMI Database (MMTCO$_2$e)

Category	2006	2007	2008	2009	2010	2011	Total	N
Total GMI reductions	16.7	22.8	43.0	40.8	41.2	38.8	203.3	542
USG-attributable reductions	12.3	15.6	34.8	29.2	28.4	25.8	146.1	532
Reductions for non-China activities	6.7	9.6	26.8	15.6	14.8	10.8	84.3	474
USG-attributable non-China reductions	5.0	6.1	23.1	10.7	9.0	4.9	58.8	464

SOURCE: "Activities" table data joined with "Emissions" table data, GMI database, as of March 19, 2012.
NOTE: The "N" in the final column refers to the total number of activities with emissions reductions.

Table 2.11
Emissions That Could Lead Directly to Emissions Reductions, by Activity Type

Activity Type	Total Activities	Activities with Emissions Reductions	Percentage of Type
Study	465	4	1.0
Project	785	535	68.0
Other	21	1	4.5
Total	1,271	540	42.0

SOURCE: "Activities" table data joined with "Emissions" table data, GMI database, as of March 19, 2012.
NOTE: There were two activities that did not have a "type" recorded.

edge more accessible across countries). In Table 2.11, we narrow the set of activities to the three types that could reduce emissions *directly*: feasibility studies, demonstration projects, and "other." There were 1,271 activities of these three types, and of these, 540 (42 percent) had actual emissions reductions recorded in the GMI database.

Table 2.12 breaks down emissions reductions by year and region. The total reductions reported in this table are slightly lower than those in Table 2.10 because some activities were classified in a way that did not allow us to associate them with a specific region.

A large number of the activities implemented between 2006 and 2011 had *potential* rather than actual reductions in emissions of methane recorded. Of the roughly 2,000 activities in the GMI database, 373 were reported to have at least some potential reductions in methane emissions.[19] This means that the activity in question involved estimating the emissions reductions that were feasible at the site in question (e.g., farm, gas facility, coal mine), although the projects required to achieve the associated emissions reductions had not necessarily been undertaken. Potential reductions reflect a substantial effort on the part of EPA and its partners to identify and quantify methane reduction opportunities, which is an important output of the GMI program and USG funding.

---

[19] Of the USG-funded activities, 305 included potential reductions.

Table 2.12
Actual USG-Attributable Emissions Reductions, by Region (MMTCO$_2$e)

Region	2006	2007	2008	2009	2010	2011
East Asia and Pacific	7.3	9.7	11.8	18.8	19.8	21.2
Europe and Central Asia	3.3	3.3	3.5	5.4	3.8	3.7
Latin America and Caribbean	0.0	0.0	0.1	1.1	2.0	0.9
Middle East and North Africa	0.0	0.0	0.0	0.0	0.0	0.0
North America	0.0	0.0	0.0	0.0	0.0	0.0
South Asia	0.0	0.0	0.0	0.0	0.0	0.0
Sub-Saharan Africa	0.0	0.0	0.0	0.0	0.0	0.0

SOURCE: "Activities" table data joined with "Emissions" tables data, GMI database, as of March 19, 2012.

Gender-Related Outcomes

DoS has recently increased its emphasis on tracking program outcomes relative to women's access to adaptation and mitigation technologies and programs related to climate change. A focus on gender is appropriate because women are more vulnerable to climate change because they often lack economic and social rights, especially in developing countries (UNDP, 2009). In post-disaster situations, women are often discriminated against during the distribution of resources and are responsible for looking after sick family members and engaging in subsistence activities (Bäthge, 2010).

One of our evaluation goals was to identify the activities and outcomes (whether partly or wholly supported) that increase access for women to adaptation and mitigation technologies and opportunities, specifically by documenting GMI contributions to the three DoS metrics that have a gender dimension:

- number of women trained in global climate change
- number of institutions serving women with an improved capacity to address climate change issues as a result of USG assistance
- number of laws, policies, agreements, or regulations addressing climate change that directly affect women that have been proposed, adopted, or implemented as a result of USG assistance.

In the GMI database, the metric that comes closest to measuring the number of women trained in global climate change is participation in GMI activities. Table 2.13 shows that, from 2006 to 2011, 2,286 women participated in GMI activities.[20] Women constituted 15 percent of all participants when we look at total participation in GMI activities. Although the GMI database contains some historical data on female participation, EPA did not start trying to comprehensively collect these data until 2011, following a request from DoS. Although the total number of women participants has been gradually increasing, the average percentage of

[20] Of this total, 94 percent participated in USG-funded activities.

Table 2.13
Female Participation in GMI Activities Recorded in the GMI Database

Category	2006	2007	2008	2009	2010	2011	Total
Participation in GMI activities	20	51	222	328	437	1,328	2,386
Participation per activity	20	6	9	11	10	23	15
Percent of total participation (%)	3	3	16	14	11	26	15
Participation in training/ workshops only	20	9	166	222	296	306	1,019

SOURCE: "Activities" table, GMI database, as of March 19, 2012.

women participating has not changed significantly—except for a sharp increase in 2011. The rise in female participation and share of all participants in 2011 is likely an artifact of the participation data collection process, given DoS and EPA's recent increased attention to gender outcomes. Of the women participating in GMI activities, the majority participated in training and workshops, as shown by the last row in Table 2.13. However, in 2011, there was a significant increase in women's participation in other activities, excluding training and workshop programs.

We were unable to provide insights with respect to the other two gender-focused metrics using quantitative data because data pertaining to these two metrics had not been collected in the GMI database. During our site visit to the Philippines, we visited agricultural demonstration projects with pig farmers, most of whom were women. We discuss the gender impacts we observed during our site visits in Chapter Three. As noted, EPA only started collected gender-specific data—at the request of DoS—in 2011, and data collection for outcomes other than participation/training is difficult because these metrics are hard to quantify. Furthermore, the relationship between the metrics and funding is not always clear. For these reasons, in Chapter Four, we provide suggestions concerning how we could think about these metrics and measure them.

Leveraged Funding

In addition to the dollars spent by the USG on GMI activities, EPA also tracks leveraged funding associated with GMI (see Table 2.14). Leveraged funding includes funding by other governments, network members, or the private sector "that augments or builds upon an activity or effort funded by the U.S."[21] Leveraged funding might include additional activities that a project host site (e.g., landfill manager, farm owner) undertakes as a result of an activity funded by GMI.

---

[21] Each year, EPA, as the lead USG agency for GMI, prepares an annual report on behalf of the United States that documents the program's accomplishments achieved as a result of U.S. activities. Part of that documentation entails collecting data on the amount of funding or resources that are leveraged by USG funding or activities. We reviewed a document titled "Methodology for Estimating Leveraged Funding for USG Activities in Support of the Global Methane Initiative" (EPA, 2011a). Although the document is not available to the general public and thus is not reproduced here, the quotes in this section are from that document.

Table 2.14
Annual USG Funding and Leveraged Funding, FYs 2006–2010 ($ millions)

Category	Fiscal Year				
	2006	**2007**	**2008**	**2009**	**2010**
Total USG funding	12.66	11.73	10.10	9.50	9.90
Leveraged Funding	25.90	9.81	6.23	109.44	11.41

SOURCE: Total USG funding amounts provided by EPA. Leveraged funding amounts were taken from the GMI database.

In preparation for its annual report on GMI, EPA counts total leveraged funding for the year. Since GMI is a voluntary program, there are no reporting requirements for member countries or Project Network members, so the leveraged funding amounts in the GMI database are not comprehensive. EPA reports that it "collects information on funding or resources that are leveraged either formally (e.g., through matching funds from an EPA-funded grant or cooperative agreement), or more informally (e.g., in-kind resources provided by a company that is hosting the site of a measurement study or pre-feasibility study)." As a result, EPA believes that the leveraged funding amounts reported substantially understate the actual funding leveraged by USG expenditures on GMI.

The leveraged funding documentation we reviewed explains how EPA estimates leveraged funding, but it does not provide information on methods. It states that "EPA collects information on funding or resources that are leveraged," but it does not provide information on how those resources are identified (systematically or unsystematically), what the parameters are for considering an effort to be leveraged funding, or how leveraged funding is estimated.

EPA states that the leveraged funding amounts "are quite conservative and significantly underestimate the actual amount of funding that is in fact leveraged based on the USG investments in GMI." This may indeed be the case. During our site visit to India, we were told about funding for the clearinghouse that was provided by the government of India, but this did not show up as leveraged funding in the documentation. According to in-country officials, $622,000 was provided for the clearinghouse by Coal India Limited on behalf of the Ministry of Coal, but this amount was not captured in the GMI database.

Alternatively, leveraged funding amounts could be overestimated. EPA cited $235 million in leveraged funding for 2004, which accounted for more than half of the total documented leveraged funding through FY 2010 ($398 million). This expenditure appears to have occurred prior to the launch of GMI. Therefore, it is hard to argue that a GMI activity was leveraged because the program did not exist. The leveraged funding data we reviewed included varying degrees of details about leveraged funding efforts and the role of a USG-sponsored activity as a catalyst to that effort. This would make it difficult to do any systematic assessment of leveraged funding and its merits.

It is also clear that not all reported leveraged funding actually constitutes "leveraged" funds in terms of contributions to methane reduction activities that occurred as a result of USG contributions. As currently defined and tracked, leveraged funding for GMI seems to conflate funding that actually *leverages* USG funds ("augments or builds upon an activity or effort funded by the U.S.") and funding for methane emissions reduction efforts from any source other than the USG. We recognize that tracking leveraged funding for a voluntary program is both very important and extremely difficult. In Chapter Four, we suggest how GMI could collect this information more systematically, even if there is no way to collect it more comprehensively.

Summary

The available quantitative data on GMI is useful for summarizing the overall scope of the program and how the program has evolved over time. Largely through USG support, approximately 2,000 activities have been completed across seven regions and four sectors. Some regions, such as Sub-Saharan Africa, South Asia, and the Middle East and North Africa, have engaged in relatively fewer activities than other regions, and emissions reductions were largely concentrated in the East Asia and Pacific region. China accounted for a substantial share of total methane emissions reductions, even though DoS funding does not support GMI activities in China. Through GMI, USG-supported training activities drew nearly 7,000 participants, although we do not have quantitative measures that capture the specific benefits that accrued to those who received training. Regarding DoS value added, we attribute approximately one-half of all outputs and outcomes to DoS funding contributions. Finally, data on leveraged funding show substantial financial contributions from other sources toward reducing methane emissions, but it is difficult to separate non-USG funding that resulted from USG support from funding that was provided in parallel to USG support.

CHAPTER THREE

Qualitative Observations of DoS Contributions

As discussed earlier, GMI is a partnership meant to promote methane capture and use activities and reduce barriers to implementing these activities, rather than solely implement emissions reduction technologies. As a result, much of its effort and value is not captured in the quantitative information outlined in Chapter Two. In an effort to collect information on GMI that was not captured in the database and to assess the value added from DoS contributions to the non-quantitative portions of GMI, we also collected qualitative information about GMI, its activities, and outcomes. Collecting this qualitative information involved examining program documentation (e.g., GMI history, organizational structure, accomplishment reports, Steering Committee and subcommittee reports), programmatic guidance in the IAAs, interviews with relevant program officers in DoS and EPA, and three country site visits that included field visits to witness GMI-implemented activities and interviews with local funding recipients and other stakeholders.

The site visits provided insights into GMI beyond what can be seen through the tally of activities and outcomes from the database. The site visits gave us information about hard-to-measure components of the program, such as how activities are implemented on the ground, what outcomes are achieved from the activities for which we do not have good quantitative outcome measures (such as meetings and workshops), and overall views of GMI from participants and stakeholders in other countries. Hence, the site visits served as the primary source of qualitative information about GMI activities and outcomes.

From the program documents and interviews, we explored how GMI functions as a program: how it is organized, how decisions are made regarding program direction, how funding decisions are made at the activity level, how these activities are implemented and monitored, and how GMI engages new partners and Project Network members. We also examined the role of DoS in these GMI functions and identified the specific and unique contributions that the department has made to the program. This constituted the main qualitative information from which we assessed the DoS value added in providing programmatic and strategic support to GMI. In this chapter, we begin by reviewing observations from the site visits regarding GMI activities and outcomes and then present observations about DoS value added in GMI programmatic activities.

Site Visits

Three RAND team members conducted site visits in three GMI member countries. Each visit lasted approximately one week, during which the team member conducted semistructured interviews with key stakeholders and field visits to GMI activity sites.

We had several goals in conducting the site visits. First, by conducting semistructured interviews with key stakeholders, we aimed to understand how activities were being implemented, what challenges were encountered by those engaged in GMI activities, and the outcomes realized. By studying on-the-ground implementation, we hoped to understand what outcomes local organizations attributed to GMI funding and support. We also expected GMI to benefit from hearing participants' views about the program and suggestions for improvement. Second, through conversations with representatives from key organizations, we wanted to identify possible spillover effects or leveraged opportunities that are often hard to quantify and therefore may not have been captured in the GMI database. We also wanted to determine the extent to which on-the-ground participants were aware of DoS involvement in GMI and the value added that might come from that involvement. Third, the site visits allowed us to collect data on specific GMI activities to supplement and validate the data available in the GMI database in an efficient way. Fourth, since DoS has recently added gender impacts to the metrics that GMI should collect, we attempted to determine those impacts in our site visit locations. We noted the number of women we encountered in our interviews and field visits and inquired about the participation of women in GMI activities during our interviews. In total, we interviewed 32 stakeholders, discussed approximately 30 distinct activities, and conducted six field visits.[1] The small number of activities reviewed relative to the number of activities represented in the database (2,000) does limit the utility of this means of validating the data.

Site Selection Process

The solicitation for this evaluation called for one to three site visits. We chose to visit three countries. The sites were selected in collaboration with EPA; given EPA's role as distributor of GMI funds and as the ASG lead, EPA staff are informed about the various GMI activities in different countries.[2] The selection was designed to maximize the utility of the site visits across several dimensions. To guide the selection process, we identified features that each individual country should have and the requirements that the set of three countries should meet as a group, to the best that we could meet these criteria with only three sites. Ultimately, we determined that each country should

- be a substantial methane emitter (in at least one sector)
- be engaged in a substantial number of GMI-funded activities
- be the site of at least some activities that were conducted prior to 2011.

---

[1] One field visit may have involved several activities. For example, each of the two field site visits in the Philippines involved paying a visit to two different farms where the demonstration projects were located.

[2] Additional information on the site selection process can be found in Appendix D, along with a discussion of the interview protocol development and the process for contacting respondents.

It was our goal that the set of three countries would be heterogeneous geographically, in terms of income, and in terms of emissions volumes; would cover the four major GMI sectors; and would represent a range of activity types.

We excluded China from the candidate countries due to restrictions on using DoS funding for projects in that country. We gave a higher weight to countries where RAND staff had language expertise or local support systems (or both). Logistical considerations also factored into final site selection. For example, were there enough relevant activities within a reasonable geographic radius that the field visits could be conducted in the approximately weeklong timeframe?[3] Given these criteria, we chose to visit India, Mexico, and the Philippines.

Data Analysis

We conducted the qualitative data analysis in two stages. In the first stage, we collected field notes pertaining to each of the three countries that we visited. We organized the data into the topical areas that we had identified earlier—namely project implementation, impacts, funding, other support from GMI and overall views about GMI. We then identified common themes and outlier cases for each area. In particular, we identified common themes pertaining to specific GMI sectors. When analyzing these data, we also paid attention to the background of the respondent (such as whether he or she worked in a public or private institution), whether the respondent or the respondent's institution had received funding or in-kind support, and the extent to which the respondent was familiar with GMI.

In the second stage, we compared the common themes and outlier cases identified earlier across all three site visits. We first wanted to determine whether particular sector characteristics were country-specific. We also wanted to determine whether observed commonalities in a given sector could be found across all three sites. The ultimate goal was to identify commonalities at the sector and country levels, respectively.

In the process of conducting interviews, we also collected quantitative data from respondents in conversations and through documents provided after the interview. We compared these data to the data in the GMI database in an effort to validate the database.

To obtain perspectives from stakeholders who were not directly engaged with GMI, we contacted environmental NGOs working in or near our selected sites. We focused on organizations that did not receive direct funding from GMI but were interested in GMI's efforts, given their own portfolio of work.[4] We were interested in learning whether these organizations knew about GMI and, if so, what their impressions were of the program and its related activities. We were keen to hear from these organizations because they could help us assess the local visibility of GMI and its work and identify spillover effects from GMI-related activities.

We developed a list of NGOs for each of the three countries by consulting two sources of information. First, we referred to the list of Project Network members on the GMI website; the network is described as consisting of private-sector entities, financial institutions, government bodies, and not-for-profit organizations, all of which are interested in reducing methane emissions and using the methane captured as fuel. Project Network members are considered actively involved in GMI (GMI, undated[b]). From this list of Project Network members for each country, we extracted a list of institutions classified as not-for-profit and NGOs. We

---

[3] We had four to five days on the ground to conduct field visits and interviews and two to three days for travel.

[4] On our final list of NGOs, only one had received direct GMI funding.

hoped to identify organizations that were familiar with GMI activities and that could speak to how successful those activities were locally or how GMI was viewed locally. We also used web searches to identify other organizations that did not self-identify with GMI to get a sense of how well known GMI might be in the region. While reviewing each organization, we read its profile information to ensure that the organization was an environmental NGO working on or concerned with methane emissions.

Once our list of NGOs was developed (25 NGOs in total), we randomly selected organizations to contact. We selected at least one NGO from each source of information for a given country. We then obtained contact information and contacted our selected list of nine NGOs both by email and telephone. The response rate was very low, however, so, overall, this effort did not contribute much to our analysis.

In the remainder of this chapter, we present our findings from the site visits. For each country that we visited, we provide a description of our site visits and views of key stakeholders as they pertained to outcomes, funding, and overall perceptions of GMI. We conclude our discussion of the site visits with a summary of the main lessons learned.

India

India is the second largest emitter of methane in the world. In 2004, the Indian government was one of the 14 national governments that were charter partners to GMI (M2M, 2009a). Since the country's involvement with GMI, organizations in India have taken a proactive role, involving themselves in all four sectors of GMI.

India has 4 million households that rely on biogas for energy. There are nearly 2,000 large-scale biogas systems in India (M2M, 2009a). In the coal sector, India has successfully established the Coal Mine Methane (CMM)/Coalbed Methane (CBM) Clearinghouse. The CMM/CBM Clearinghouse was established through the collaborative efforts of India's Ministry of Coal and GMI. It is a key contact for both domestic and international organizations interested in the development of coal mine and coal bed methane projects in India. India co-chairs its coal subcommittee and, as of this writing, five CMM tenders were being floated for commercialization.

In the landfill sector, five pre-feasibility studies exploring the economic viability of LFG have been completed. The LFG project at the Gorai Landfill is already in operation. In the oil and gas sector, the Oil and Natural Gas Corporation (ONGC), which dominates India's crude oil and natural gas production industry, has reduced methane emissions by approximately 10 million metric standard cubic meters (MMSCM) since joining GMI in 2007.[5] India hosted the 2010 Methane to Markets Partnership Expo in New Delhi, which attracted about 500 participants from around the world, including private, not-for-profit, and government organizations. GMI describes the expo as a very successful, highly visible event that involved significant intragovernmental coordination. Some of those we interviewed from the coal sector identified significant positive outcomes from attending the event.

---

[5] These data were provided to us by one of our interviewees working in the oil and gas sector in India.

Site Visit Description

Ongoing successful efforts in India, as well as the immense potential it holds for reducing methane emissions, made it a logical candidate for site visits. Among the four GMI sectors, we focused on the coal and the oil and gas sectors because most of the activities conducted under GMI in India were in these two areas, though we also explored some landfill activity. We interviewed key stakeholders in the landfill sector and those engaged with cross-sector activities. We were interested in cross-sector activities that focused on information dissemination, primarily because this sector generates outcomes that are often hard to measure quantitatively and are best captured and understood qualitatively through conversations.

Table 3.1 provides a summary of our site visits in India. Interviews were conducted over a period of eight days and ranged in length from half an hour to more than half a day, depend-

Table 3.1
India Site Visit Summary

Site Visit Participant	Sector	Main GMI Activities Discussed	Type of Data Collection	Received Direct Grant Funding for Implementation Activities?	Received Support for Other Activities?	Number of Key Interviewees
Coal advisor	Coal	Interactions and involvement with GMI	Interviews	No	Yes	1
Central Mine Planning and Design Institute	Coal	Clearinghouse	Site visit and interviews	Yes	Yes	2
Government institute	Coal	Feasibility studies, workshop	Interviews	Yes	Yes	1
Consulting organization	Landfill	Interactions and involvement with GMI	Interviews	Yes	Yes	1
Landfill in western India	Landfill	Site visit to a medium-sized landfill	Site visit and interviews	No	Yes	1
Landfill in metropolis	Landfill	Interactions and involvement with GMI	Interviews	No	No	1
ONGC	Oil and gas	Interactions and involvement with GMI, measurement studies, training programs, study tours	Interviews	No	Yes	2
Oil and gas facility in western India	Oil and gas	Site visit to a medium-sized oil and gas facility	Site visit and interviews	No	Yes	1
Not-for-profit organization	Cross-sector	Workshops, expositions, study tours	Interviews	Yes	Yes	1
Diplomatic mission	Cross-sector	Interactions and involvement with GMI	Interviews	No	No	0

NOTE: The category "Received Funding for Other Activities" includes USG funds provided for such activities as conferences, study tours, and training programs.

ing on the respondent's availability. In addition to interviews, we also conducted two site visits: to a landfill and to an oil and gas facility. Because we wanted to a capture a diverse range of views and experiences, we interviewed personnel working in different geographical regions and with varying levels of involvement with GMI in terms of funding and other support received.

In the tables in this chapter, the categories "Received Direct Grant Funding for Implementation Activities" and "Received Support for Other Activities" were created based on interviewee responses. In the latter case, the USG funds multiple activities, such as conferences, study tours, and workshops. According to interviewees in all three countries, USG funding levels for such activities are variable, and most could not place a dollar amount on the cost of these activities. Respondents from the coal sector provided an example: When coal officials attended study tours in the United States, their tickets were paid for by the organization, while all the on-the-ground logistics expenditures were paid for with USG funds. These respondents did not know how much funding was used for these activities, either from their own organizations or from the USG.

Next, we highlight the key themes that emerged from our conversations.

Impacts

During our site visits, we discussed impacts pertaining to each of the respective sectors. We focused specifically on those impacts that were not been captured by quantitative measures but that were highlighted by respondents as being highly important.

Oil and Gas Sector

In the oil and gas sector, the most evident and easily measured impact is methane emissions reductions. Participants mentioned several efforts undertaken as part of GMI that led to direct reductions in methane emissions—for instance, through directed inspection and maintenance, technological interventions, and the adoption of the identification, quantification, and reduction (IQR) methodology. Implementation of a tank vapor recovery system at one plant resulted in methane reductions of 2,025 cubic meters per day. Replacement of the servo gas system with instrument air helped reduce output by 850 cubic meters of methane emissions daily. From these efforts and others, respondents from the oil and gas sector estimated that there had been methane reductions of approximately 10 MMSCM since India joined the partnership in 2007, equivalent to reducing 139,000 $MTCO_2e$ from the atmosphere. They felt that such outcomes stemmed from opportunities to learn about new technologies and measurement studies conducted under the aegis of GMI.

In addition to reductions in methane emissions, respondents mentioned several outcomes that are more difficult to quantify. After the implementation of the IQR methodology in the oil and gas sector, more than 350 leaks from 56 installations were identified and mended. Interviewees noted that fixing these leaks not only improved operational efficiency but also resulted in improvements to the environment and the safety of workers at their oil and gas plants. GMI supported an effort by enlisting HY-BON Engineering, a Texas-based firm, to train a core team in India's oil and gas industry in IQR techniques between June 25 and July 1, 2009.

Respondents felt that they benefitted most from learning about new technologies during study tours, training programs, or conferences. For instance, oil and gas personnel mentioned that they learned about the use of infrared cameras for measuring leaks when GMI contracted with consultants that used these cameras for this purpose. Since then, respondents had purchased and used their own cameras. They also noted that their organization had its own GMI operational team and was taking a structured approach to implementing GMI activities. Due to their organization's success with GMI, they felt that their company had become a success story for others in their industry. Gas Authority of India is one of the latest companies to express an interest in participating with GMI in India.

Coal Mining Sector

The CBM/CMM Clearinghouse is a key contact for both domestic and international organizations interested in coal mine and coal bed methane projects in India. This organization provides much-needed information to investors to promote the country's CBM/CMM market. The CBM/CMM Clearinghouse was established in November 2006 through a memorandum of understanding between GMI and India's Ministry of Coal. GMI provided the clearinghouse with initial funding for a three-year period (personal communication, January 12, 2012).

Since the establishment of this clearinghouse, large numbers of commercial CMM projects have been developed in India. Respondents involved with managing coal mining projects felt that participation in the 2008 workshop "Development of CBM/CMM in India: An Opportunity Area" encouraged them and their organization to identify CMM projects for commercial development in an environmentally beneficial manner. According to one interviewee, "The offering of the CMM blocks is the result of a collaborative effort between [coal agencies] and EPA's Coalbed Methane Outreach Program." Another group of coal mine managers noted that by presenting their work in a workshop that was organized under the GMI umbrella, they were able to establish networks, especially with private mining companies. As of this writing, these officials were in talks with some of these companies to conduct feasibility studies in other mining fields. Interviewees from two different organizations in the coal mining sector remarked that the potential of coal mine methane has generated great interest among Indian companies that have begun to enter this field.

Coal mining managers mentioned that they benefited from activities that built in-house capacity. By attending the Coal Mine Methane Conference in Alabama in 2010, coal officials learned about emissions-related regulations that they felt would help them provide guidance for developing national policies. They were of the opinion that study tours and training programs organized under GMI provided much-needed experience to officials responsible for teams working in CBM/CMM/VAM (ventilation air methane) development in India. Along with human capital, one respondent also noted acquiring physical capital under a GMI grant, which in the future would allow his organization to conduct feasibility studies independently.

While these benefits accrued to those directly engaged with GMI, other people and organizations have benefited from GMI activities that have brought the issue of methane emissions and the opportunities in this field to their attention. The GMI India and the CBM/CMM Clearinghouse websites serve as examples of activities that have created a platform for information-sharing and dissemination. Those who have benefited from this knowledge have taken steps to share it. Coal-sector officials reported providing training to 14 students from the Indian School of Mines to expose them to new technologies that were not being taught in the school.

Landfill Sector

As of this writing, five pre-feasibility studies had been conducted in India's landfill sector, and one LFG project in one landfill had materialized from these studies. When respondents were asked to discuss benefits, all cited the case of the Gorai Landfill in Mumbai. The success of methane extraction from the landfill has earned the city $5.2 million in carbon credits from captured gas, 70 percent of which was methane. This LFG project is expected to provide energy to the local community (Methane International, 2009). Respondents did not identify other quantifiable benefits in addition to this case.

Overall, respondents felt that making key stakeholders aware of the potential of extracting gas from landfills provided the greatest benefit from involvement with GMI. According to a respondent working in the landfill sector, "The issue of methane extraction from landfills has seen the light of day due to EPA efforts. It has brought awareness to stakeholders." Interviewees noted that providing information on the potential benefits of methane recovery and utilization has encouraged new organizations, such as municipal corporations and private organizations, to explore landfill gas-to-energy projects. Most of these less quantifiable benefits were cited by respondents from private organizations.

Cross-Sector Activities

In the case of cross-sector activities, respondents found conferences to be useful networking opportunities. Conferences also permitted them to learn about the latest developments in their fields in leading countries. While attending the 2010 Methane to Markets Partnership Expo in New Delhi, respondents from the coal sector mentioned that they took the opportunity to interact and deliberate with one of the leading VAM technology providers in the world. The opportunity to meet such a contact was invaluable because he helped them prepare a VAM project that could be funded by the Government of India. One interviewee saw such interactions as opportunities that pushed his company to explore new areas.

Gender Impacts

During our site visits in India, we did not receive feedback about gender impacts stemming from the efforts being conducted under the aegis of GMI. In India, we were mostly focused on the coal and oil and gas sectors, industries in which women are usually underrepresented. Even when we asked respondents to discuss gender impacts, most mentioned that this question did not apply to them because women do not work in those industries in India.

Funding

In all our interviews, none of the respondents expressed problems with the funding process. They mentioned that they had received funding through one of the following two mechanisms: They either responded to a call for proposals or received a subcontract based on their expertise. Funding levels provided by GMI ranged from 100 percent for all activities outlined in the contract to small-scale financial support for conducting study tours and workshops. Irrespective of the funding mechanism and the funding levels received, respondents unanimously stated that communications with EPA personnel about funding were always excellent and that they always received funds in a timely manner.[6]

The importance of funding to respondents depended on the nature of their activities, as well as the sector in which they were engaged. For instance, for personnel working on cross-sector activities, funding was extremely important because it enabled them to extend their

---

[6]   EPA personnel distribute the funds for GMI activities and therefore are the people who are most frequently in touch with stakeholders in partner countries.

outreach: "Our work in providing sensitization is critical. However, we need to take it to the next level. For this, funding is critical." Respondents felt that funding opportunities in India from private and government bodies were limited; thus, they viewed funding for cross-sector activities from external sources as very beneficial. Representatives from large organizations with substantial funding of their own noted that funding was not the core motivator in their involvement with GMI. Respondents from such organizations said that their relationships with GMI were invaluable because of the opportunities to learn about new technologies and share information.

Views of GMI and Suggestions for Improvement

All respondents voiced favorable opinions about the program. They did not mention any major structural weaknesses in GMI, although they voiced preferences and opinions regarding where the program should be heading.

Several respondents reiterated the benefit of information-sharing and having the opportunity to learn about new technologies. One described GMI as a "good collaborative platform where one company can see the work of others." He stated that since all the information is pooled in one place and all associated organizations, such as companies and vendors, gather in the same place, GMI provides a structured portal for bringing together information and organizations. Another respondent saw GMI as a "storehouse of knowledge" and "central point of contact" that ensured that all member countries were on the same page.

While all respondents appeared satisfied with their involvement in GMI, they provided several suggestions for improvement. Several respondents felt that GMI should shift its focus. Suggestions differed depending on the sector. One respondent felt that GMI needed to conduct more feasibility studies and engage more with industry, especially in the agricultural sector in India. This sector consists of large numbers of small farmers and businesses that would not be able to engage in methane collection on their own. The respondent felt that there should be more funding for feasibility studies and opportunities for technology transfer.

One respondent in the landfill sector expressed a desire to move from pre-feasibility studies to actual demonstration projects. He felt that, in India, private companies saw the potential for methane to energy in the landfill sector but were afraid to commit money. According to the respondent, entrepreneurs in India need to see successful demonstrations before entering this market. He recommended that GMI invest in more demonstration projects in the landfill sector to encourage growth. In line with these two recommendations, one of our respondents stressed the need to shift focus from coal, oil, and gas to the agriculture, livestock and landfill sectors, where the potential for methane reduction is much higher. To convince ministries in these sectors of the potential of these activities, one participant stressed the need to set up a GMI office in India. He was of the opinion that an on-the-ground presence would help move things forward. In the coal sector, respondents requested more focus on VAM, carbon capture and storage, and CMM recovery.

These recommendations stressed shifting funding priorities. A few recommendations were directed at the structure of the program. One respondent who received funding talked of the difficulties of having to pay 100 percent of the money up front to purchase equipment but only receiving funding in tranches. He suggested that GMI should make exceptions to their payment schedules, depending on the needs of the project. Respondents requested more activities involving technology transfers. They recommended that once personnel were trained in India, they should be used to train personnel in other South Asian countries. Doing so would

---

GMI in India: Oil and Natural Gas Corporation (ONGC) India

ONGC is a state-owned Indian oil and natural gas company and one of the largest oil and gas exploration and production companies in Asia. Since its involvement with GMI, this company has been involved in several technology transfer workshops, measurement studies, pre-feasibility studies, and study tours. Through various efforts conducted in collaboration with GMI, ONGC has been able to reduce emissions by approximately 10 million metric standard cubic meters since joining the partnership in 2007, equivalent to reducing 139,000 $MTCO_2e$ from the atmosphere. The most notable feature of this company is that it has undertaken a structured approach to implementing GMI program initiatives. ONGC has an in-house operational core GMI team that surveys ONGC sites for methane emissions, uses appropriate technical interventions to reduce emissions, and monitors and reports yearly on emissions of methane. ONGC also highlights GMI activities in its annual report submitted to EPA and in its monthly newsletter. Through this newsletter, we found in our field visits that most of ONGC's on-the-ground staff were aware of GMI and the various efforts being undertake in conjunction with it.

---

not only reduce costs, but it would also create incentives for personnel in India to become better trained in technologies to reduce methane emissions because these individuals would have an opportunity to showcase their talent.

Mexico

As of 2009, Mexico ranked sixth in the world in terms of global methane emissions (184.82 $MMTCO_2e$). Since its involvement with GMI, various efforts undertaken under the program have targeted 61 percent of Mexico's total methane emissions (M2M, 2009a). LFG feasibility studies have also been conducted at four landfills. Mexico co-chairs GMI's oil and gas subcommittee. Several measurement studies have been conducted for PEMEX (Petróleos Mexicanos), the state-owned company that is in charge of gas and oil production in Mexico. In the coal mining sector, there have been demonstration projects, as well as changes to proposed regulations that would allow coal mine methane recovery. The Mexican government is working to create a national program to capture methane gas emissions from animal waste (M2M, 2009a). Mexico has also hosted several GMI subcommittee meetings and workshops, including the ministerial meeting in Mexico City on October 1, 2010, at which GMI was launched. Domestically, the Mexican government has started a program similar to GMI known as the National Strategy on Climate Change that focuses more broadly on mitigation measures for climate change (M2M, 2009a).

Site Visit Description

Because of the extent to which Mexico is involved in reducing emissions of methane and for reasons of geographical diversity (our other two site visits were in Asia), we selected Mexico as our second site visit. Mexico was also closely involved with the initiation of both GMI and its predecessor, Methane to Markets. Moreover, because it is such a large emitter of methane, Mexico has great potential to curb emissions. In Mexico's case, we focused on oil and gas and landfills, the two sectors into which most of the GMI-sponsored activities have tended to fall. Because PEMEX is the only company in Mexico's oil and gas sector, we talked to PEMEX employees who had been involved with GMI. In the case of the landfill sector, we interviewed five stakeholders involved with various landfills in the north and southeastern regions of the country. We visited one landfill site in southeastern Mexico.

Table 3.2 shows the range of sectors and activities we covered in Mexico. Interviews lasted from one hour to a few hours. All interviews were conducted over a span of five days. We focused on the oil and gas and landfill sectors (Table 3.2). To capture a range of opinions and experiences within this sector, we interviewed both private and public officials working in different regions of the country who have had varying levels of involvement with GMI.

Impacts
Oil and Gas Sector

According to interviewees from PEMEX, the company engaged in efforts to reduce methane emissions as a result of its involvement with GMI. An experienced contractor that had USAID funding conducted methane emissions measurement studies at a PEMEX plant. The results led the company to replace wet seals with dry seals in gas compressors, which significantly reduced methane emissions from gas venting. One respondent said that the success of these efforts had made many in his company realize the potential economic and environmental benefits of emissions reduction initiatives. At the time of the site visit, PEMEX was conducting similar measurement studies with technical assistance from an EPA team. The interviewee considered the role of GMI to be critical in initiating and helping PEMEX continue its methane reduction efforts. He believed that ongoing support would be critical in ensuring the continuation of such efforts.

In addition to measurement studies, PEMEX has also been involved with GMI through chairing its oil and gas subcommittee. By organizing and participating in conferences, one key respondent felt that PEMEX had greatly benefited from learning and sharing information with other companies in partner countries.

Table 3.2
Mexico Site Visit Summary

Site Visit Participant	Sector	Main GMI Activities Discussed	Type of Data Collection	Received Direct Grant Funding for Implementation Activities?	Received Support for Other Activities?	Number of Key Interviewees
PEMEX	Oil and gas	Measurement studies	Interviews	No	Yes	2
Large landfill in metropolis	Landfill	Pre-feasibility study, guidance on developing requests for proposals	Site visit and interviews	No	Yes	1
Conglomerate	Landfill	Study tour, technical assistance	Interviews	No	Yes	1
Municipality in southern Mexico	Landfill	Workshop, pre-feasibility study	Interviews	No	Yes	1
Nonprofit organization	Landfill	Implementation of training program, technical assistance	Interviews	Yes	No	2

NOTE: The category "Received Funding for Other Activities" includes USG funds provided for such activities as conferences, study tours, and training programs.

Landfill Sector

Methane emissions from landfills in Mexico have yet to be significantly reduced because only one of the LFG projects was in operation at the end of 2011. However, in our interviews, respondents gave examples of several efforts that they believed would result in reductions in methane emissions in the future; in the case of one of the landfills in the southeastern region of the country, GMI had conducted a pre-feasibility study. Respondents affiliated with the landfill sector and working in different regions in Mexico noted that municipalities often face the challenge of not knowing how to structure a technically complicated bidding process in accordance with Mexican law. Given this constraint, GMI is on the right path by funding a research project that will provide guidelines on how to conduct a bid for technical services to reduce methane emissions from landfills. The impact of this activity is still unknown.

In addition to its own involvement in providing technical expertise, GMI has contracted with an NGO working in Mexico to provide technical assistance and guidance to officials in the southeastern region of the country. The NGO is developing a master terms of reference (TOR) document, which will function as a guidebook for municipalities wanting to engage in LFG projects. According to a representative of the NGO, this form of assistance is much needed to move the landfill sector forward because most municipal governments in Mexico stay in power for only three years. Having a TOR guidebook to build from would also give municipalities more time to engage in bidding processes instead of having to directly recruit companies that may not provide them with the best service.

Municipal officials provided examples of how they benefited from receiving technical support. According to one of the municipal officials at a landfill in southeastern Mexico, the help provided under GMI had been essential because people like him have received guidance on what type of generators are needed and what price factors they should be using in the bid: "They [EPA personnel] know which technologies will be available in the near future, and that is very important because it determines how you write the Terms of Reference," he said. Respondents particularly valued the nonprofit character of EPA, which encouraged them to trust EPA's advice. GMI is providing much-needed exposure and guidance to public officials in Mexico who consider this support necessary as they embark on the process.

Given this support, most respondents were optimistic about reductions in methane emissions from landfills in the near future. One of the municipal officials working in the southeast asserted that the landfill in his municipality would close sometime in 2012, after which the feasibility study conducted under GMI would be used to implement an LFG project. In our own research, we found that after one landfill had been closed in Mexico, the Federal District published a declaration of need to hire an entity that could capture and exploit the biogas. We saw this publication as a positive sign of efforts being undertaken to implement LFG projects in Mexico. If an LFG project were to be implemented at this landfill site, one public official noted, there would be an estimated reduction of 2 $MMTCO_2e$. Respondents were also aware that such projects could earn them carbon credits under the Kyoto Protocol, so there were also potential financial benefits to the projects.

Other efforts under GMI were also indirectly contributing to reductions in methane emissions. GMI organized a study tour of LFG projects in the United States for representatives from Mexican corporations. The study tour induced a large transnational private company to consider implementing LFG projects. As an official from that company who attended the study tour noted, "Getting to see what plants in the U.S. were doing was crucial." Having received technical advice from EPA personnel, this organization has been able to find potential sites for

LFG projects. Personnel interviewed believed that their organization's involvement with GMI has provided it with much-needed exposure and guidance.

While some of these efforts have had direct impacts on reducing methane emissions, other efforts have had spillover effects. Municipal officials working in the southeastern region of the country organized a workshop with EPA and the Mexico's Ministry of Environment. The officials not only benefited from learning about methods of reducing methane emissions, but they also had an opportunity to network. The workshop helped them establish relationships with their peers in other municipalities and open up lines of communication with them. Along with such opportunities to build or rebuild relationships, municipal officials also expected to benefit from receiving training on how to conduct the bidding process for LFG projects. The nonprofit organization mentioned earlier is expected to host this workshop after it has developed the master TOR document.

Gender Impacts

Similar to our experience in India, respondents did not have much to say about the gender impacts associated with GMI because the program focuses on industries that do not tend to attract women. Only two of our key respondents were women. They were working in the oil and gas and landfill sectors, but they did not discuss any gender issues. The only substantive information we received regarding gender was that seven out of 40 participants in a GMI-organized workshop concerning landfills were women.

Funding

With the exception of the NGO that was funded to conduct the research project and workshop mentioned earlier, all other organizations received in-kind support. Respondents from PEMEX thought that GMI intervention was extremely important. They felt that public organizations like theirs set their yearly goals based on visible outcomes, such as numbers of barrels of oil or the quantity of gas produced. Given these types of metrics, respondents said that it was often hard to convince company decisionmakers to invest in methane emissions reduction programs, especially before the economic benefits of those measures have been demonstrated. For these reasons, respondents were thankful that the first measurement studies were conducted with external funding and guidance. Otherwise, these studies would have not happened, and the follow-up actions (such as the change in seals) would not have occurred. They also mentioned that because the funding did not have to go through the recipient's budget (i.e., the measurements were conducted by consultants paid directly by EPA or USAID), the project was able to proceed without the complications of using internal funding, such as internal regulations and bureaucracy.

Some respondents provided suggestions for improving the funding process. Respondents working in the oil and gas sector preferred to keep GMI funds away from the government bureaucracy because they feared that if the funds had to go through government channels, nothing would be accomplished. The NGO was the only entity that had received direct GMI funding. A representative from this organization felt that although the funding process was not onerous, staff had a difficult time filling out the forms. He recommended that GMI provide support to help answer questions and suggested that making the process easier could attract more proposals.

Views of GMI and Suggestions for Improvement

All respondents had favorable opinions of GMI. Public officials were impressed with the "zero-bureaucracy" nature of the program. They felt that they had been able to undertake projects that they would not have otherwise because the application and distribution process was so streamlined. Projects also started much faster; they were able to begin ventures before the economic value had been demonstrated.

Respondents appreciated receiving sound technical advice from an impartial source. Public-sector officials, especially in the landfill sector, spoke of their distrust of advice from private companies. They felt that such companies highlighted only the technologies that they want to sell and that government officials had no way to judge the value of their advice. These officials trusted advice from officials working for GMI because they know that GMI does not have a commercial interest in the technologies offered. The respondents observed that personnel working for GMI are always abreast of the latest technologies and provide "solid-gold" advice.

Several respondents working both in the private and public sectors spoke highly about the quality of service they received under GMI. When respondents from a private company that is undertaking LFG projects were asked about the strengths of GMI, they compared it with private consulting and engineering firms with which they had existing relationships. They chose to work with GMI because it is experienced with the subject matter. GMI personnel work quickly and are sensitive to time constraints. According to one respondent, "It doesn't affect [the company] that they [EPA] are a public institution. They work as efficiently as a private institution." Public officials in the southeast compared the advice they received from GMI to some of the work they were conducting in collaboration with an aid agency from another country. In both cases, they felt that they were getting advice of roughly equal quality. But they felt that their communications with GMI were much better and that the relationship was closer.

When asked to discuss what features can be improved in the operations of GMI, public-sector officials working in the landfill sector in the southeast felt that GMI needed to have more personnel spending more time on the ground to get things moving locally. They also felt that when future workshops are organized, GMI should also invite local speakers because such speakers would be more conversant on local issues and could provide relevant advice to the audience.

Other participants spoke about the structural weaknesses within Mexico that prevented them from moving forward with projects. They felt that the greatest hurdle was the internal bureaucracy of their own organizations. The law that regulates purchasing makes it difficult for them to buy equipment. Without this equipment, they are unable to conduct measurement studies on their own. In the absence of such measurements, they cannot make arguments to their board that substantial amounts of methane are being emitted and that reducing methane emissions means that they could sell those credits through the Clean Development Mechanism (CDM). In the landfill sector, respondents discussed challenges with registering LFG projects under the CDM, which they felt was vital if their projects were to be become economically viable.[7] They believed that GMI-related individuals were likely familiar with the process

---

[7]   The Clean Development Mechanism is defined in Article 12 of the Kyoto Protocol. It allows an Annex B country to implement an emissions reduction project and earn certified emissions reduction credits (United Nations Framework Convention on Climate Change, undated).

of selling carbon credits through the CDM and that any advice they could provide would be useful.

The Philippines

The Philippines joined GMI in 2008 after first learning of the program from a U.S. embassy official serving in Manila.[8] GMI began funding activities within the country in 2009. The Philippines ranks 35th in global methane emissions, which stem primarily from the agricultural and landfill sectors. GMI projects in the Philippines have focused primarily on agriculture, but there was one landfill project as of this writing. GMI's in-country investment began with the development of the Resource Assessment for Livestock and Agro-Industrial Wastes–Philippines (International Institute for Energy Conservation, Eastern Research Group, and PA Consulting Group, 2009). The resource assessment provided a strategic framework for methane reduction activities in the Philippines.

This resource assessment identified methane emissions from individual subcategories of the agricultural sector (pig farming, coconut processing, alcohol distilleries, and slaughterhouses). The assessment found that pig farming had the highest potential for methane reduction and carbon offsets. In the Philippines, about 30 percent of hogs are raised on commercial farms and 70 percent on small farms. Commercial farms have to comply with effluent requirements for waste, but small farms are exempt. In its efforts in support of GMI, the USG (EPA) chose to focus its methane reduction activities on small farms.

Site Visit Description

The Philippines was selected as one of our site visits because of its activities in the agricultural sector, its geographic location, and similar methane reduction activities in nearby countries (e.g., Vietnam). EPA, in support of GMI, began working with Philippine government agencies to develop an approach to engage backyard farmers by educating and training them on the use of methane capture technologies and the economic benefits of using the captured gas in the household, such as for cooking and lighting.

Our site visit included interviews with the primary government agencies that had been engaged with EPA, institutions that finance emissions reduction projects, and two agricultural cooperatives, since EPA, in support of GMI, is working with existing cooperatives as a means of engaging backyard farmers.

Table 3.3 provides a summary of our site visits and interviews in the Philippines. We conducted semistructured interviews and visited several small farms that had received biogas installations as part of GMI activities.

Impacts

The Philippines has a strong network of agricultural cooperatives (co-ops). Co-ops consist of individual local farmers who join together to realize economies of scale by jointly purchasing supplies, such as seeds or animal feed, and services, such as veterinarian care. EPA decided to

---

[8] A U.S. embassy official first approached an undersecretary at the Department of Science and Technology about the prospect of the Philippines joining GMI. The undersecretary attended a couple of GMI-related meetings, and a partnership was established.

Table 3.3
Philippines Site Visit Summary

Site Visit Participant	Sector	Main GMI Activities Discussed	Type of Data Collection	Received Direct Grant Funding for Implementation Activities?	Received Support for Other Activities?	Total Number of Key Interviewees
Cooperative south of Manila	Agriculture	Attended training workshops; installation of tube-bag biodigesters on individual farms	Site visit and interviews	No	Yes	1
Cooperative north of Manila	Agriculture	Installation of tube-bag biodigesters on individual farms	Site visit and interviews	No	Yes	2
Development Agency of the Philippines	Agriculture	Conducted training workshops	Interviews	No	Yes	1
Department of Science and Technology	Agriculture	Assisted with training workshops	Interviews	Yes	Yes	3
National Solid Waste Management Committee	Agriculture	Development of the Philippine Methane Initiative	Interviews	No	No	2
Landbank and the World Bank	Agriculture	Attended training workshops to help cooperatives apply for financing of methane reduction activities	Interviews	No	Yes	4
Large-scale farmer	Agriculture	Attended training workshops	Interviews	No	No	1

NOTE: The category "Received Funding for Other Activities" includes USG funds provided for such activities as conferences, study tours, and training programs.

work through the existing agricultural co-ops as a way to engage small farmers in methane reduction activities. This approach makes sense, as co-ops have the necessary relationships with individual farmers and are in a position to educate, train, and monitor methane recovery projects.

Training and Education Workshops

In support of GMI, EPA awarded a grant to the Philippines' Department of Science and Technology (DOST) to provide a training program for technicians, who could then be deployed to work with the individual co-ops to install methane recovery technologies. This effort is referred to as a "train-the-trainers" workshop. Representatives from GMI and other experts provided the technological knowledge on how to educate farmers and install and manage the technologies. DOST engaged the Development Academy of the Philippines (DAP) to assist with the workshops, because DAP has the expertise to develop training materials and provide training programs.

DOST, with the help of DAP, held three additional workshops in the major pig-farming regions of Luzon, Visayas, and Mindanao. These workshops were supported by USG funds.

Technicians who were trained in the initial workshop now train and educate small farmers and local authorities about the potential for methane recovery as it relates to pork production. EPA has engaged at least two agricultural co-ops, which have installed tube-bag biodigesters at some small farms to serve as demonstration projects and training sites.

EPA, in support of GMI, also engaged the World Bank and Landbank, which acts as the World Bank's coordinating and managing entity for its CDM activities in the Philippines. Landbank participated in the training workshops for pig farmers in Luzon, Visayas, and Mindanao. It has lending centers and development assistance centers in each province, so its employees know many of the farmers and were able to identify potential attendees for the workshops. Landbank employees also helped develop content for the workshops.

Landbank was able to use EPA's resource assessment to create a program of activities related to the conversion of animal waste to energy. The resulting carbon credits can be used as part of the CDM, and pig farms that sign up for the program are eligible to receive income from the sale of the credits. Landbank helps their operations comply with the necessary rules and regulations and helps them obtain financing related to the installation of the digesters. Participating in the program also makes the farms eligible for technical and capacity-building assistance from EPA.

As a capacity-building activity, Landbank and EPA, in support of GMI, also created a manual for measuring methane collection. Landbank is also developing a program of activities related to capturing landfill methane emissions in partnership with EPA. EPA will assist with developing measures for calculating landfill methane reductions.

Demonstration Projects

In addition to meeting with representatives from the agencies mentioned earlier, we visited two of the agricultural co-ops with which EPA has partnered. These co-ops are demonstration sites for tube-bag biodigester systems.

The first co-op visited was in the Batangas area, south of Manila. The co-op learned about GMI from one of Landbank's field personnel from the local Landbank Development Assistance Center. The timing of the partnership was good for the co-op, as it had recently decided to expand its operations but was struggling with how to comply with both the Clean Air Act and the Clean Water Act of the Philippines. These laws were designed to address the smell, noise, and waste of pig farming. The use of digesters mitigates these problems, so the installation of biodigesters helped the co-op solve a problem that could have hindered its plans for growth.

There are currently 100 pig farmers in the cooperative, and the majority of those farmers are women. EPA, through GMI, provided the equipment and supplies for fabricating tube-bag biodigesters. Members of the co-op, with technical support from the Bureau of Animal Industry (BAI) and EPA, fabricated the digesters and installed them at four farms as a means to pilot the effort. We visited two of those farms. The pig farmers, both women, were happy with the digesters and with the fact that the captured gas is stored onsite. They used it primarily as fuel for cooking and lighting. At one of the farms, the cook stove was used by three families. The co-op asks farmers to document their meter readings everyday so that it can track the amount of gas being produced and used.

Prior to using a biodigester, households typically purchased canisters of liquefied natural gas as an energy source. The gas from the biodigesters saves the equivalent of 1 canister of liquefied natural gas, which typically costs $16 a month. Some of the households were still using firewood as cooking fuel. For those homes, the methane provides a much cleaner source of cooking fuel. The availability of methane also alleviates the need for women to spend time finding and cutting firewood. As a result of these demonstration projects, many more farmers in the co-op are interested in having the biodigesters installed. EPA has agreed to provide the supplies needed to fabricate 44 more biodigesters for this co-op. At the time of our visit, the co-op was raising the money and support to fabricate and install the biodigesters.

These demonstration projects seemed to be serving their purpose. The co-op has had inquiries from a co-op from another municipality that is interested in acquiring digesters for its own pig farmers. As a result, the co-op we visited was in discussions with the International Training Center for Pig Husbandry about forming a partnership to conduct training activities for other co-ops. The center has the capacity to develop training materials and conduct training courses; the co-op does not. It also has field personnel throughout the Philippines and could transfer both the knowledge and the technology to interested co-ops.

The second co-op we visited was located north of Manila in Bulacan. It has 800 members, 300 of whom are pig farmers. The majority of the pig farms (about 200) are small farms, and 13 of these have biodigesters. The co-op owns the piglets, provides the feed, and provides veterinary services; in other words, the co-op engages in contract animal husbandry. The farmer applies for a loan, and the coop then builds the pen. Each farmer starts with about 12 piglets.

DOST regional personnel connected the Bulacan co-op with EPA. EPA is providing a grant for the equipment and supplies needed to build the tube-bag biodigesters. The co-op will charge "at cost" for the biodigester on a payment plan. The co-op will then use that money to buy more supplies for more farmers. The co-op will be trained to install both the tube-bag digesters and the dome digesters. BAI and DOST will provide technical support for the first few months after installation. After that time, the co-op is expected to provide support, since its members will have been trained by BAI and DOST.

EPA support was meant to jump start these methane capture and use activities at the co-ops. Co-op personnel indicated that, in the future, Landbank will provide loans to the co-op to purchase equipment and supplies. The co-ops will hire a third-party financial adviser to help them determine their financing needs, develop a business plan, and conduct feasibility studies. The local Landbank Lending Center and Development Assistance Center will then work with the co-op on its business plan, which will estimate the number of digesters to be installed per year, the rate of absorption, and so on. The local DOST staff who were working with this co-op thought that after the business plan was developed, Landbank would have the confidence to loan money to the co-ops because BAI and DOST had been involved. Those two agencies are involved as a direct result of EPA support and activities. The Bulacan co-op would like to see all 300 pig farmers install digesters, since it will help the co-op manage waste and comply with air and water regulations. In turn, the co-op hoped that the digesters would help entice more members to become pig farmers.

Philippine Methane Initiative

At the national level, involvement with GMI has encouraged several government agencies to set up a country-level equivalent, the Philippine Methane Initiative (PMI). Those interviewed from government agencies recognized that GMI activities have been a good starting point but that it is time for the Philippine government to step in and provide the support for ongoing monitoring and evaluation, as well as to develop a national strategic plan for methane capture and recovery across the Philippines. These officials observed that getting national agencies to take over these functions is the best way to make these activities sustainable over time. EPA, in support of GMI, has provided funding for the preparatory work and strategic planning necessary for developing the program. PMI will be under the authority of the National Solid Waste Management Commission and will address both agriculture and landfill issues. The main agencies to be involved are DOST, Landbank, DAP, and the Department of Agriculture, which houses BAI. It is not clear what the role of GMI will be once PMI is set up. Interviewees envisioned that GMI would continue to introduce PMI to new techniques and technologies for methane capture and recovery.

When the Philippines joined GMI in 2008, approximately 300 biogas systems had been installed throughout the country under previous methane capture initiatives (see International Institute for Energy Conservation, Eastern Research Group, and PA Consulting Group, 2009). DOST maintains a database of these biogas systems, although it is not clear how many of them are still operational, as that information was not maintained over time. These initiatives often fell short in the areas of maintenance of the equipment and monitoring of the emissions. Proponents of PMI hope that a national initiative can focus on maintenance and monitoring, as well as documentation of projects.

Gender Impacts

The Philippines' engagement with GMI is only a few years old. The focus on agricultural co-ops thus far has tended to involve a fairly good representation of women, from what we witnessed. As was mentioned, many of the pig farmers who participate in the co-ops are women, and most of the demonstration projects on the farms that we visited were run by women. Pig farming is primarily done by the women as a means of bringing in additional income. Their husbands often hold jobs outside of the home or are crop farmers. Interestingly, at the two co-ops we visited, it was mainly women who worked in the administrative offices as well. As methane capture and use activities in the Philippines move into other areas, such as landfills, we expect the number of women involved in these new initiatives to be quite small, as few women work in that industry. It is also interesting to note that we met with government officials in four different government offices, and about half of the representatives with whom we met were women.

Funding

In total, in support of GMI, the USG has spent $775,000 on activities in the Philippines since 2009. Approximately 43 percent of that was spent on Philippines-based agencies, universities, and consultants, with the remainder provided to U.S.-based contractors. The activities supported ranged from strategic assessments, such as the Resource Assessment for Livestock and Agro-Industrial Wastes, on-the-ground training workshops, and demonstration projects, to feasibility studies and physical equipment and supplies.

Those we interviewed did not raise significant concerns about funding, in terms of either the length of time to receive funding or the related reporting requirements. The resources appear to have acted as a catalyst and have coalesced the interests of several national agencies. From our interviews, it appeared that these agencies often contributed in-kind resources in terms of staff time for planning and participating in the workshops, as well as co-op and interagency engagement. Interviewees viewed participation in GMI as a synergistic opportunity to benefit from the reduction of GHGs, which is a national goal, and introduce sustainable processes and, potentially, develop and sell carbon credits.

Views of GMI and Suggestions for Improvement

Those we interviewed felt that participation in GMI activities had been beneficial. EPA has planted the seeds for methane reduction in the Philippines, and Filipinos have been trained in best practices by knowledgeable people in the field and have developed good contacts for further technical assistance. EPA has also helped develop standards and regulations and has been helpful in teaching Filipinos about financing options for emissions reduction efforts. Further GHG reductions are expected, since capture and recovery efforts are relatively new in the country.

Those with whom we met in related Philippine government agencies conveyed the importance of understanding local customs and social interactions in making these types of programs successful. Several government partners had experienced difficulty in working with a project manager because of a lack of cultural understanding and appreciation for local government protocol. This problem was mitigated by bringing on a respected Filipino engineering professor and long-time USAID consultant. The consultant was familiar with the government agencies and their employees, was able to successfully negotiate relationships and activities, and interacted with the Philippine partners on an ongoing basis. The partners in the Philippines were well organized and coordinated as a result. The consultant's ability to facilitate these interactions was likely instrumental in the movement to develop PMI.

Interviewees had some suggestions for how to improve the process for holding GMI workshops and training sessions. The agencies that had received funding to hold such workshops in the past observed that a longer planning horizon would go a long way to developing more effective workshops. Organizations were often only given a few weeks' notice to plan a workshop. Based on their experience, having several months to plan these workshops would allow them to be more successful in targeting the right type of attendees and would provide a better chance of convincing invitees to attend. Potential attendees often have to request travel time and funds and need time to make such arrangements.

In our discussions with Landbank, interviewees said that they needed a formal agreement with EPA for their planning purposes. Most government agencies, including Landbank, need to provide justification when they request travel funds. These individuals could refer to a formal agreement when they request travel to attend a GMI-related activity or in spending time preparing workshop materials. The formal document would also allow them to do more long-term planning with EPA. Previous efforts with EPA have been on very short notice and often difficult to accommodate, especially in budgeting for travel.

Feedback from Nongovernmental Organizations

We contacted nine NGOs working in the vicinity of all three of our field sites. Despite making three attempts to contact these organizations by both email and telephone, we heard back from only three, one in the Philippines and two in India. Although all three NGOs had heard about GMI, only two of them were familiar with GMI activities.

The NGOs that were familiar with GMI seemed to have only a cursory knowledge of the program. Employees of the NGO in the Philippines mentioned that their organization had consulted the GMI website on occasion over the past four years and had joined the GMI network in 2011. Their interest in GMI stemmed from their participation in the promotion and construction of biogas digesters in the Philippines. The NGO in India was aware of GMI's work because it engages with policymakers at the federal and state levels and with technology providers who are conducting activities in collaboration with GMI.

Only interviewees at the Indian NGO talked about the benefits of GMI. They said they had benefited from attending the exposition in New Delhi, where they learned about the various on-the-ground activities being conducted in conjunction with GMI in India. Their assessment was that the CMM sector is suffering from regulations that hinder the implementation of methane collection technologies.

Observations from Site Visits

Due to the small sample size (three countries), it would be inappropriate to extrapolate site visit observations to GMI and DoS contributions as a whole. Although we tried to select representative partner countries, there are factors specific to each member country that influence its experience with the program. Nevertheless, we believe that interaction with on-the-ground GMI participants was useful in helping us understand how the program is being implemented, what factors are working to make the program a success, and what shortcomings need attention.

Knowledge of DoS Involvement Was Limited

During our site visits, we asked whether respondents knew of the involvement of DoS in GMI efforts and what benefits or challenges they identified with GMI. Across all three countries, respondents were aware of the involvement of a U.S. government agency, but only in the Philippines were they aware of DoS involvement specifically. As mentioned earlier, it was a U.S. embassy official who first introduced GMI to Philippine government officials. Financial officials from the World Bank and Landbank in the Philippines favorably weighed the fact that DoS was a partner. Nongovernment officials, such as members of the agricultural co-ops implementing methane capture devices, were not aware of DoS involvement. They primarily associated GMI with EPA, which was expected, given that the technical assistance and funding come from EPA on behalf of USG contributions to GMI. Many respondents replied that, for them, it was only important that a U.S. government agency was involved because this would give their work more credibility in the domestic market.

GMI Is Increasing the Momentum for Methane Reduction Activities

Across all three sites, respondents working in different sectors were in agreement that one of the biggest advantages of being involved with GMI is that it brought to the forefront the benefits associated with reducing methane emissions and the potential to use methane as a fuel. In the Philippines, two of the government agencies, DOST and BAI, had been previously involved in methane capture initiatives in the agricultural sector. However, earlier initiatives seemed to last only as long as the technologies functioned properly. Once methane capture devices began to fail, there was no backup system or technical expertise to ensure that the equipment continued to operate. To this end, GMI resources appear to have acted as a catalyst that coalesced the interest of several national agencies and local stakeholders to build a support framework that will keep biodigesters operating under the technical oversight of DOST and BAI. In India, respondents in both the landfill and coal sectors mentioned that GMI had increased the momentum of methane reduction activities.

Several respondents appreciated the global nature of GMI. Cross-sector activities, such as conferences, gave respondents a forum to learn and network with others working in their sector. Organizations and their representatives gave high marks to study tours and training programs that helped build in-house capacity in their organizations. In the Philippines, government technicians were being trained to construct and maintain biodigester technologies so that the technologies could be transferred more widely throughout the country. In Mexico, municipal officials received advice and information from workshops about LFG projects in their localities. In India, study tours for coal, oil, and gas officials provided valuable information about new technologies. In turn, these officials shared their knowledge with others, either in their organizations or locally. The impacts of capacity-building, knowledge-sharing, and transfer and networking opportunities are often hard to capture. However, such cross-sector activities have proved to be important catalysts of activity in our three site visit countries.

Capacity-building activities have helped institutionalize methane-reducing practices in the Philippines. In that country, involvement with GMI has encouraged several national agencies to establish a country-level equivalent, the Philippine Methane Initiative. The aim of PMI is to develop a nationwide strategic plan for methane recovery and capture. This includes building in monitoring and assessment functions. In India, ONGC had identified a core team to survey ONGC sites for methane emissions, use appropriate technical interventions to reduce methane emissions, and engage in yearly monitoring and reporting of methane emissions. These examples show how those benefiting from involvement with GMI are taking a structured approach to implementing GMI activities. Engaging local organizations to take initiative and responsibility is perhaps the best way to move forward for GMI.

Local Presence and Demonstration Efforts Have Been Effective

After our site visits, we identified areas that would benefit from further attention. In comparing and contrasting the three countries, we found that an on-the-ground presence for GMI may play an important role, especially in sectors that have several small stakeholders. For instance, in the Philippines, the local GMI representative was able to successfully negotiate relationships and activities among government employees and private individuals. His involvement appears to have provided a stable source of coordination that proved beneficial for promoting a national effort. On the other hand, in India, some respondents observed that a local presence was critical, especially for the agricultural and landfill sectors, which consist of many smaller entities. Coordination issues were not mentioned in the coal and oil and gas sectors, because these sectors in India are dominated by a few large firms. For these reasons, we feel that GMI would benefit from having local representatives, especially in fragmented sectors, to help coordinate activities and engage government officials.

The importance of funding differed, depending on the nature of the sector and activities. Funding was deemed most important by those engaged in cross-sector activities and in the agricultural sector, which consists of many small entities. Those engaged in cross-sector activities, such as organizing workshops or study tours and spreading information, felt that, without GMI, they would have struggled to obtain funding from domestic sources and likely would not have succeeded. Leveraged funding was more important for activities conducted by government officials and state companies. In Mexico, respondents from PEMEX noted that the first measurement studies would not have taken place without external funding from GMI. Successful demonstration projects are necessary for some companies to make investments in such activities. Similar views were expressed by respondents in the landfill sector in India. They stated that private companies were unwilling to take the risk of investing in LFG projects without seeing an actual, functioning project. For these reasons, in industries dominated by small private firms, demonstration projects are likely to be necessary to even have a chance of transforming those industries. Demonstration projects may also have the potential for large payoffs in terms of transferring the technology.

DoS Value Added in GMI Programmatic Activities

In evaluating DoS value added in the context of GMI's programmatic activities, we considered how DoS contributed to the establishment of GMI and how it participates in GMI's ongoing operations. What specific skills or expertise are unique to DoS and have been used to the benefit of the GMI partnership? In this section, we describe these important and specific contributions.

Establishment of GMI

According to our interviews with DoS and EPA staff who were engaged with GMI, DoS played a substantive role in the establishment of GMI. EPA brought the necessary technical expertise but according to our interviews with EPA and DoS staff involved at the time, a good portion of the foundational work (diplomatic and administrative) for establishing the international partnership was performed by DoS. DoS had experience with establishing similar international partnerships and so was able to apply that knowledge to writing the underpinning documents, such as GMI's TOR, laying out the organizational structure, the responsibility of the partners, and the functions of the partnership. Our interviews also revealed that DoS was pivotal in using its influence and diplomatic competence to engage new countries as potential GMI partners, as well as to encourage existing partners to become more actively involved. Each country presented its own challenges and competing interests, so specific knowledge of how to tailor an appeal based on the interests of that country was important for success.

DoS was also able to apply its specific set of skills in providing some training and guidance for the subcommittee leadership and members when GMI was first established. At that time, many of the subcommittee members and some leaders did not have experience with participating in or running international meetings that needed to build consensus and make recommendations. The DoS program officer at the time attended the subcommittee meetings and provided guidance on how to chair a meeting in an international setting, how to interact with participants, how the participants could effectively interact with each other, how to motivate discussion and ideas to reach consensus, and how participants could replicate the process in their own countries.

Programmatic and Strategic Guidance

DoS provides ongoing support and direction to GMI through its continued efforts to promote GMI to potential new partner countries and through its participation in the Steering Committee and the ASG. DoS described its process of using numerous tools to promote GMI over the years, such as educating embassy staff about the opportunities to engage in GMI in the countries where they are serving. A successful example of this outreach was described in the Philippines case study, where Philippine government officials were first introduced to GMI through staff from the U.S. embassy.

A recent example of DoS efforts to educate staff who may come in contact with potential partner countries took place prior to the October 2011 all-partnership meeting. The DoS program officer hosted a briefing on the program and its benefits with the environment and economic officers of country desks to educate them about GMI. The program officer asked them to raise the issue at any bilateral meetings with potential partner countries and to encourage attendance at the all-partnership meeting. About 20 country desk officers attended the briefing.

As a member of the Steering Committee, DoS brings its knowledge of individual countries, its diplomatic skill, and its responsibility for promoting U.S. foreign policy when considering the approach and direction of GMI. From what we heard in our interviews with DoS and EPA staff, DoS brings a unique perspective and skill set that differ from those of EPA, the other U.S. representative on the Steering Committee.

DoS also participates in the ASG. EPA points out that although DoS participation may be less frequent than the day-to-day functions carried out by EPA in its roles on the ASG, DoS brings a special set of skills to the group. Its ability to engage with other countries using its various diplomatic tools allows it to be effective and strategic in gaining the attention of other countries. These skills have been used to engage partner countries (existing partners as well as potential new partners) through official discussions and cables.

CHAPTER FOUR

Findings and Recommendations

The primary purpose of this evaluation was to identify the value added of DoS contributions to GMI in FYs 2006–2010. To do this, we structured an analysis that first examined the overall GMI program at an aggregate level against which we could measure DOS value added. To assess DoS value added, we examined DoS contributions to GMI (funding as well as programmatic and strategic support) and looked at both quantitative and qualitative output and outcome measures, including information gathered from three country site visits.

The previous two chapters described the detailed quantitative and qualitative information collected as a result of that effort. In this chapter, we summarize the overall key findings from that examination. We also present some recommendations for how data collection could be improved to answer more sophisticated questions in the future about the effectiveness of GMI and the value added by DoS contributions.

Key Findings

DoS Funding for GMI Has Been Substantial

As an international program to promote voluntary reductions in methane emissions, GMI has increased its membership from 14 country partners in 2004 to 41 in 2011. This suggests that GMI's approach to methane emissions reduction and reuse has appeal as a cost-effective approach to reducing GHG emissions and attracting partner countries. DoS has played an important and significant role in USG support for GMI by providing more than half of the program's total funding since FY 2006.

Through DoS support, GMI has conducted a variety of activities (about 2,000) that range from training local governments and the private sector about methane reduction opportunities and technologies (about 6,900 people had been trained as of this writing) to implementing demonstration projects that others can see and then adopt (146 MMTCO$_2$e of methane emissions reduced as a result of USG support).

DoS Has Supplied Strategic and Programmatic Support

DoS has provided leadership support since the program was created. It was instrumental in establishing and shaping the partnership by crafting the charter documents that established the multicountry agreement, and the department has used its diplomatic and policy skills to work with partner countries and to attract new members. DoS has continued its support through its membership on the Steering Committee, which provides strategic and programmatic direction to GMI. It also supports the ASG periodically by facilitating diplomatic engagement with partner countries.

Site Visits Suggest That GMI Activities Are Seeding Methane Reduction Efforts

From our site visits in India, Mexico, and the Philippines, we were able to capture stakeholders' views on the role GMI plays on the ground. In most of our interviews, respondents who worked across all four GMI sectors said that they had benefited from capacity-building activities, such as study tours, conferences, and workshops that provided exposure to new ideas and increased their skill sets. Information and networking opportunities allowed respondents to explore new markets, learn about new technologies, and, in some instances, build in-house capacity to implement methane reduction projects. While the GMI database lacks information on "softer" outcomes related to informational and networking opportunities that are often hard to quantify, these outcomes were the ones most commonly cited during our site visits.

Many of our interviewees noted that GMI has increased people's awareness of the benefits associated with reducing methane emissions and the potential to use methane as a fuel. In India, respondents working in both the landfill and coal sectors mentioned that GMI had increased the momentum for pursuing methane reduction activities. In the Philippines, government technicians were being trained to construct and maintain biodigester technologies. In turn, they held multiple sessions to train others so that the technologies could be transferred more widely throughout the country. In Mexico, USG-supported measurement studies in the oil and gas sector conducted at one plant identified methane leaks. The national oil company is now replicating those measurement studies at its other plants to identify leaks.

Recommendations

We conclude by identifying opportunities to support and improve GMI program implementation and enhance DoS's ability to assess progress in the future. We note the need to solicit feedback from project participants and to enhance GMI data collection, especially to support future program evaluation. We also discuss potential improvements in evaluation metrics and identify new ways to conduct future program evaluation. Although the recommendations are founded on the data and evidence that we collected, they necessarily reflect our subjective judgment as evaluators.

Soliciting Feedback from Project Participants

Based on interviews during our site visits, we recommend the USG (EPA and/or DoS) solicit feedback from local stakeholders more frequently. In our interviews, we found that local stakeholders were aware of problems in implementing projects, but felt that they did not have sufficient avenues through which to convey these observations to the USG. DoS should consider supporting this process, especially in soliciting feedback from government-affiliated entities or by working with EPA to enhance the channels through which stakeholders can provide information that will help improve GMI.

GMI Database

The GMI database is an important tool for GMI and one that EPA has worked hard to develop. The database was created primarily to help EPA track GMI funding and activities and to record outcomes. It has a straightforward online interface that is accessible from any location with Internet access, which is important for EPA staff and contractors working in the field.

However, there are two important limitations to the database. The first involves data entry consistency and completeness: many different people enter information into the GMI database, and they do not all enter it in the same way. This leads to gaps in data coverage and conflicting conventions for how information is recorded, undermining quality and consistency. Although data gaps are less critical for EPA staff, who are intimately familiar with GMI program details, they make data use challenging for external or new users. Fortunately, this is a relatively straightforward issue to resolve through better database documentation, user guidance, and systematic data reviews and revisions—all of which EPA has begun undertaking. Over the course of our evaluation, EPA made substantial progress in "cleaning up" the database and clarifying data entry procedures.

The second issue, which is more challenging, is that the GMI database is not strictly designed to support program evaluation. While it contains many important pieces of data, such as emissions reductions or participation in GMI activities, the database structure limits the kinds of analysis that can be undertaken. For example, it is possible to break down actual emissions reductions by sector, year, or activity type. However, because of the way in which activities are matched to funding vehicles, it is not straightforward to calculate USG dollars spent per activity. This means that important evaluation metrics that could inform decision-making (e.g., emissions reduced per dollar spent on different types of activities) cannot be calculated using GMI data in their current form. The lack of funding details for a large subset of all activities in the database is another limitation. Closer examination of individual database entries showed that USG funds were used in some capacity for the activities with missing data; thus, the lack of funding information made it difficult to accurately assess GMI contributions that could be linked to financial support.

EPA has indicated that the inclusion of an activity in the GMI database reflects that there was some USG involvement in that activity, even if there was no direct USG funding; consequently, funding may not be the best indicator of USG involvement. Given the USG's prominent and long-standing role in methane emissions reduction programs, this assumption is reasonable. However, it should be documented better. To accurately assess the benefits associated with USG financial support for GMI, EPA and DoS should consider developing criteria by which they can be credited with involvement for non–USG-funded activities and then attempt to collect those data for inclusion in the database, similar to the attribution factor approach used for methane emissions reductions. This process would be challenging, but it would allow EPA, DoS, other USG entities, and other GMI participants to assess USG contributions in a clear and transparent way.

Tracking GMI Emissions Reductions

As EPA acknowledges, GMI's voluntary and international nature makes robust data collection challenging. In particular, it is difficult to track all emissions reductions associated with GMI, and it is difficult to determine how much credit USG agencies should receive as a result of their work to foster GMI activities. Some of these challenges could be reduced by making emissions accounting methodology documents more detailed and applying them consistently. Based on our review of EPA's methodology documents, we offer the following suggestions for improving emissions tracking.

- *Tiered credit versus full credit.* It is notable and commendable that EPA recognizes that not all activities should receive equal credit for emissions reductions, but the attribution factor approach could be more uniformly applied across the various sectors. Two of the methodology documents explicitly included attribution factor tiers that varied depending on the role USG support played in an activity (i.e., 90 percent, 70 percent, and 40 percent). However, the oil and gas sector methodology only allows for attribution of 100 percent of emissions reductions. In the agriculture sector, for which the methodology document is still being completed, all activities receive a 100 percent attribution factor. While there may be historical or practical reasons for these assumptions, the methodology documents currently do not explain in sufficient detail how attribution factors are developed or how they are applied.
- *Direct versus indirect emissions.* Of the methodology documents, only the one for the landfill sector specifically identifies the role of GMI activities in generating "indirect" emissions reductions—those associated with displaced consumption of other fossil fuels when combusting methane. The landfill sector is assumed to yield an indirect benefit in terms of avoided net emissions of CO_2 relative to the non-renewable, fossil-based fuels that the captured landfill gas displaces. Other sectors, such as coal or oil and gas, are not attributed with indirect emissions reductions, however. If flared methane at an oil and gas facility is captured and then used to produce energy, there may be net GHG emissions reductions from using relatively cleaner methane to generate energy. Substituting natural gas for coal, for example, would have an additional benefit above and beyond the emissions avoided by halting flaring. Total avoided emissions from activities in the coal and the oil and gas sectors may not be substantial, but by not accounting for avoided emissions across all sectors, GMI is underestimating total GHG emissions reductions.

Although we suggest that emissions tracking and attribution methodologies be improved, making these improvements could be costly. The current attribution factor tiers appear to be based on accumulated knowledge—that is, they are informed rules of thumb. Although this creates challenges for program evaluation, the current approach may be sufficient for GMI's needs. Unlike emissions reductions tracked through CDM, GMI reductions *as measured by GMI* are not being used for market transactions. Consequently, the benefits of a more complex or explicit measurement approach could be outweighed by the costs of constructing and disseminating rigorous emissions accounting procedures. To the extent that GMI wants to integrate more with programs like CDM or to the extent that the USG (including DoS) desires more certainty about GMI methane reductions, revised methodologies could be warranted.

Assessing the Evaluation Metrics

The evaluation metrics that we reviewed for this evaluation, which were defined by DoS and EPA, vary in how they align with GMI goals and in how difficult they are to measure.[1] Metrics such as *emissions* reduced are relatively easy to measure and closely align with the program's goals. Other metrics, like *capacity built*, also closely align with GMI's goals but are more difficult to measure. Some of the gender metrics do not necessarily align well, based on our conversations with a range of GMI stakeholders. Because measurement often drives what a program

---

[1] DoS applies required, standard indicators to GMI, since it falls under the "Clean, Productive Environment" program area.

focuses on, relatively weak alignment between metrics and program objectives can potentially distort performance.

Overly narrow metrics, with corresponding annual targets, may result in funding being driven toward projects that "count," such as training programs for women, rather than on efforts focused on education, knowledge transfer, or partnership-building, which may have a greater effect on the long-term goal of reducing methane emissions. Overly broad metrics may reward "quantity" rather than "quality." This is particularly important when incorporating a strong gender focus, which needs to be considered in the broader context of GMI and its goals. Forcing inappropriate benchmarks can lead to unintended effects, while not measuring progress that is desired may lead to a failure to replicate successful programs.

There is also some question as to whether the current metrics are intended to capture *program delivery* or *results*. For instance, DoS and EPA could choose to measure the number of women enrolled in training programs (an output indicator) or the number of women who have measurably gained skills through training (an outcome indicator). The latter is clearly much harder to measure, but it may be a better gauge of achievement.

The metrics as defined are fairly broad, leaving room for different interpretations by parties involved in managing and implementing GMI. Measuring the number of women trained is fairly straightforward. However, a wide range of more or less stringent criteria can be developed to classifying an organization as "serving women," or laws, policies, agreements, and regulations that "directly affect women." As a practical matter, GMI partners or grant recipients have collected minimal, if any, information on service delivery broken out by gender. These are all considerations for both DoS and EPA to deliberate as they move forward with the program. Any future evaluation of the gender impacts of GMI would benefit from more specificity as to which metrics related to gender would be of most interest to DoS and which are most applicable to GMI's goals and activities.

Leveraged Funding

Leveraged funding is an important potential benefit of DoS support for GMI. By taking a leadership role, DoS and EPA can encourage other public- and private-sector entities to increase their financial contributions to efforts to reduce methane emissions. But leveraged funding is challenging to measure. It is not clear what USG leveraged funding (as currently captured in the GMI database) includes. Leveraged funding information appears to conflate funding that "leverages" USG funds ("augments or builds upon an activity or effort funded by the U.S.," as stated in EPA's leveraged funding methodology) and funding from any source other than the USG that supports methane reductions. We acknowledge that, given DoS and EPA's substantial involvement in GMI and global efforts to reduce methane more broadly, many non-USG sources of funding *could* be interpreted as leveraged based on overall USG effort. But clearer definitions and criteria would lend additional credibility to leveraged funding estimates.

One concrete step would be to expand and clarify the methodology EPA uses to estimate leveraged funding, which currently provides little information as to how leveraged funding is calculated. The methodology states that "EPA collects information on funding or resources that are leveraged" without providing details on how those resources are identified (systematically or unsystematically), what the parameters are for deeming financial support to be leveraged funding, or how the amount of leveraged funding is estimated (i.e., Does EPA take credit for less than 100 percent of the funding amount depending on the effort, as with the emissions accounting methodologies?).

Our understanding is that the leveraged funding amounts are reported or discovered in a case-by-case manner. The footnote in the methodology we examined explains that, in preparation for reporting on GMI's first five years, EPA engaged in extensive outreach to identify investments by other countries and partners on methane recovery and use efforts under the auspices of GMI. The document indicates that these amounts were identified through the Steering Committee or sector-specific subcommittees. A similar exercise, perhaps with a lower level of effort, could be conducted on an annual basis through an online survey tool or surveys administered at all-partnership meetings. While a survey might not result in comprehensive reporting, it could be used to move to a more standardized approach that could be consistently replicated across years.

How and why a funding amount gets credited as leveraged funding could be made more explicit. Ideally, specific factors would be used to assess whether a candidate funding amount is leveraged on an activity funded by the United States. Since GMI has many partner countries and network partners who might be making such determinations, EPA should establish a common set of criteria or parameters that a funding amount should meet if it is to be categorized as *leveraged*.

EPA could also develop some guidelines for determining how much of a leveraged funding amount is attributable to U.S.-funded GMI activities. All leveraged funding might be 100-percent attributable because it is too difficult to identify the portion that resulted from a USG-funded activity, but a standard reporting guideline would help ensure clarity and consistency.

If some leveraged funding is meant to connote funding from any source other than the U.S. government, then it could be productive to create a separate category called "non–U.S. government GMI funding," rather than *leveraged funding*. The current wording *leveraged* suggests something more specific, like amplification.

We understand that there may be no way to systematically capture all leveraged funding amounts, but we believe that standards for how leveraged funding is identified, what constitutes leveraged funding, and how much of the funding is designated as leveraged would lend credibility to leveraged funding reports.

DoS in a Supporting Role

DoS has provided substantial funding to support GMI, and it has made important strategic and programmatic contributions. Although DoS provides guidance on funding allocations through the steering committee and IAA process, it defers to EPA to play the leading role in managing how USG financial support for GMI is allocated (e.g., across countries and sectors) based on EPA's technical expertise. We view this flexibility as beneficial to the program. DoS and EPA appear to have built a good working relationship, and we recommend that DoS continue to provide high-level guidance and support while allowing EPA to drive the process of identifying technical opportunities and guiding USG funding allocations to the maximum extent feasible.

Opportunities for Future Program Evaluation

Periodic *ex post* evaluations that involve a broad assessment of GMI with modest new data collection, such as this one, are important opportunities to take stock of GMI's accomplishments and DoS value added. But there are limitations to this type of evaluation in terms of depth and the ability to assess program impacts. In our evaluation, we were able to conduct site visits in only three countries; there was no scope for new quantitative data collection. Going forward, DoS (and EPA) could consider implementing a monitoring and evaluation strategy that provides greater insight into what aspects of GMI are most effectively achieving the desired program outcomes.

Impact evaluations are resource-intensive, and the scope or intensity of future evaluations should be balanced against resource and time constraints. If quantifying outputs and outcomes to track basic program performance is sufficient, then a focus on desk reviews and output measures (drawing on a more complete GMI database) would likely be timely and efficient. However, if the goal is to identify promising programs for replication or to calculate the causal impact of USG support for GMI on reductions in methane emissions, carrying out comprehensive site visits or an in-depth (and costly) impact evaluation may be more appropriate.

A true impact evaluation would also be methodologically challenging. Measuring outcomes and attributing causality is significantly more conceptually and logistically complex than measuring outputs. Ideally, an impact assessment compares changes in outcomes in the presence of a program against the counterfactual of no program. In many instances, objective measures of knowledge and skills may not be readily available, and pre-program data are likely to be nonexistent. Reasonable comparison groups may or may not be available. Finally, impacts may be inherently difficult to measure within the timeframe given for evaluation.

Based on our assessment of DoS support for GMI, we identified three activities that could supplement a long-term evaluation strategy and provide near-term insights into GMI's effectiveness, potentially at relatively low cost.

Implement a Survey of GMI Practitioners

GMI is predicated on the idea that reducing barriers to knowledge of cost-effective means of reducing methane emissions is critical to addressing global climate change. Our site visits suggest that organizations around the world view EPA as a significant source of knowledge based on its substantial technical expertise. However, there is currently no systematic way to assess knowledge transfer and capacity-building associated with GMI. A relatively low-cost way to fill this gap would be to conduct targeted surveys of individuals participating in GMI activities to assess the program's hard-to-quantify benefits. The survey could be administered annually at the all-partnership meeting to members of the public and private sectors. A short survey would allow DoS and EPA to assess (1) the types of benefits that GMI stakeholders perceive to be most valuable and (2) the types and extent of information that stakeholders gain through participating in GMI activities.

Although a simple survey would not provide rigorous estimates of capacity-building or knowledge transfer, it would provide insight into perceptions of GMI's contributions in these areas. Slightly more complex surveys could move much closer to causal estimates. A survey of participants at the all-partnership meeting could be administered twice—once before the meeting and once after the meeting—allowing DoS and EPA to measure value added directly.[2] DoS or EPA could administer surveys to a randomly selected sample of individuals working in the GMI sectors, including those with no previous exposure to GMI. Conducted annually, this type of survey would provide information on how GMI stakeholder's views about capacity-building and knowledge transfer change over time. It would also allow a comparison between

---

[2] The same type of survey could be conducted for other GMI activities, such as training or workshops, although substantial use of surveys would be costly and potentially burdensome to participants.

individuals or organizations that have worked with GMI and without prior experiences working with GMI.[3]

Use Grant Data to Assess GMI Funding Impacts

The grant-funding process that EPA administers to support methane emissions reductions is based on a competitive funding model. This model may provide an opportunity to assess the degree to which USG funding for GMI is pivotal for groups seeking support for emissions reduction activities. As EPA explained to the RAND evaluation team, grant applications are scored based on the criteria outlined in the request for proposals. Applicants that score high enough are funded, while lower-scoring applications do not receive funding. It may be possible to track unfunded grants to see whether these grants eventually receive funding from other sources. By comparing grant applications just above and just below the cutoff (for which we can assume the inherent application quality is similar), we can assess the impact of USG financial support.[4] This exercise could identify the topics for which USG support is pivotal versus topics for which there are other funding sources, allowing EPA and DoS to better target their resources.

Construct an Explicit Logic Model to Support Process Evaluation

Logic models are a straightforward way to link a program's inputs, outputs, and outcomes and to identify barriers and opportunities to effective program implementation.[5] Although EPA and DoS staff have implicitly constructed a logic model for USG support for GMI, writing down an explicit model would ensure that all staff members who are working to implement the program have a consistent view of how USG funding leads to outcomes.[6] We believe a logic model would be particularly useful in the case of USG support for GMI, given the joint role that DoS and EPA play in funding and implementing the program.

A logic model is a necessary step to facilitate a process evaluation of GMI that examines whether its activities and outputs are in line with its mission and are helping the program to reach its goals. A process evaluation would explore the efficiency of the steps involved in implementing GMI. The evaluation might include looking at the subcommittee structure (i.e., Is this the best way to develop priorities?), the process for deciding what projects and grants get funded, how funding is received and distributed by GMI, and what results are achieved.

---

[3] This is a form of what is typically referred to as a difference-in-differences analysis, which can provide causal estimates of program impacts.

[4] A version of this approach is referred to as the regression discontinuity method, although, in this case, the analysis might need to be qualitative in nature because of the relatively small sample of grant applications.

[5] Logic models can be used for both strategic planning purposes and for program and process evaluation (see Greenfield, Williams, and Eiseman, 2006).

[6] The implicit logic model was clear from our discussions with EPA staff, in particular.

APPENDIX A
GMI Partner Countries

The year each country joined GMI is listed in parentheses, followed by its regional affiliation: East Asia and the Pacific (EAP), Eastern Europe and Central Asia (ECA), Latin America and Caribbean (LAC), Middle East and North Africa (MENA), North America (NA), South Asia (SA), and Sub-Saharan Africa (SSA).

Argentina (2004, LAC)

Australia (2004, EAP)

Brazil (2004, LAC)

Bulgaria (2009, ECA)

Canada (2005, NA)

Chile (2009, LAC)

China (2004, EAP)

Colombia (2004, LAC)

Dominican Republic (2009, LAC)

Ecuador (2006, LAC)

Ethiopia (2010, SSA)

European Commission (2007, ECA)

Finland (2008, ECA)

Georgia (2009, ECA)

Germany (2006, ECA)

Ghana (2010, SSA)

India (2004, SA)

Indonesia (2010, EAP)

Italy (2004, ECA)

Japan (2004, EAP)

Jordan (2011, MENA)

Kazakhstan (2008, ECA)

Mexico (2004, LAC)

Mongolia (2008, EAP)

Nicaragua (2010, LAC)

Nigeria (2004, SSA)

Norway (2011, ECA)

Pakistan (2008, SA)

Peru (2010, LAC)

The Philippines (2008, EAP)

Poland (2007, ECA)

Republic of Korea (2005, EAP)

Russia (2004, ECA)

Serbia (2010, ECA)

Sri Lanka (2011, SA)

Thailand (2008 EAP)

Turkey (2010, ECA)

Ukraine (2004, ECA)

United Kingdom (2004, ECA)

United States (2004, NA)

Vietnam (2007, EAP)

Site Visit Interview Protocol

Name of respondent:

Position:

Organization:

Location:

GMI activity discussed:

Sector that the project comes under:

Date of interview:

Interviewer name:

I would like to start the interview by asking you to discuss the project you completed or attempted to complete under GMI. Please walk me through the various stages of the project cycle, from the time you heard about GMI until now.

Project Specifications
Note to interviewer: Please ask the respondent the following questions concerning project specifications if the respondent did not answer them earlier.

1. What led you to participate in GMI and/or apply for funding?
2. When was the project initiated?
3. How was the project executed?
4. When was the project completed?
5. What was your role in the activity?
6. Were there participants other than your organization, and, if so, what were the various roles?

Impacts
If the project has been completed:

1. How much has your project reduced methane emissions? *(Probe for a numerical answer.)*
2. How did you measure methane emissions?
3. How accurate are your measurements? *(Probe for why the respondent finds the measurement to be accurate or inaccurate)*
   a. Very accurate
   b. Somewhat accurate
   c. Not accurate

4. Were there other ways to calculate the emissions reductions and, if so, why did you choose the method you just described?
5. What other benefits did the project you completed provide? *(Probe for any gender impact.)*

If the project has not been completed:

1. How much do you expect your project will reduce methane emissions? *(Probe for a numerical answer.)*
2. How did you build that estimate?
3. How do you plan to measure methane emissions?
4. How accurate do you think your measurements will be? *(Probe for why the respondent finds the measurement to be accurate or inaccurate.)*
    a. Very accurate
    b. Somewhat accurate
    c. Not accurate
5. What other benefits do you expect your project to provide? *(These might include softer impacts that are hard to measure, such as improved awareness. Probe for any gender impact.)*

Funding

1. How was the project funded? If you applied for funding, how did you find your experience with the funding application process? *(Probe for why the respondent selected the answer he or she did.)*
    a. Easy
    b. Somewhat easy
    c. Difficult
    d. Very difficult
2. What percentage of your costs was covered by GMI funding? *(Probe for a numerical answer.)*
3. What were the other sources of funding that helped you cover your costs?
4. How could the funding process be improved?
5. Would you seek GMI funding again if you were eligible? *(Probe for why the respondent selected the answer he or she did.)*
    a. Yes
    b. No
6. Were you concerned about revealing proprietary information when participating in GMI? What was done to assuage those concerns?

Other Support from GMI

1. In addition to funding, did you receive other kinds of support from GMI during project implementation? *(Probe for whether the respondent received any of the following support features: advice on how to characterize and measure methane emissions, cost-effective measures to reduce emissions, develop best management practices, help remove barriers, or conduct feasibility studies or technology demonstrations.)*

2. Have you participated in other GMI activities? *(Probe for conferences and trainings.)*
   a. Yes *(List activities mentioned)*
   b. No
3. Of all the support you received, what do you consider the most important/valuable activities in GMI?
4. What other kinds of support would you have liked to receive from GMI?

Overall Opinions About GMI

Now, we would like to ask you about your overall impressions about GMI.

1. What, in your opinion, is the strategic idea behind GMI?
2. Do you think GMI is currently meeting its stated goals? *(Probe for reasoning.)*
   a. Yes
   b. No
3. Do you know entities or organizations in [country] that have either worked with GMI or are aware of it?
   a. Are any of them critical of GMI's work? What do they criticize?
   b. Who have you communicated with about your GMI project?
4. In general, do you have concerns about the way GMI works with partners in your country? *(Probe for reasoning.)*
   a. Yes
   b. No
5. To what extent do you feel GMI support was critical for your organization's efforts to reduce methane emissions? *(On a scale of 1–5, with 1 being not important and 5 being very important; why?)*
6. Are there other programs that you work with that provide similar benefits to GMI or have similar goals, or is GMI truly unique? What are they, and how have you worked with them?
7. Were you aware that the U.S. Department of State is a prime funder of GMI activities? If so, does its involvement in the program increase GMI's presence or validity in your eyes or in the eyes of your country?

Overall Strengths and Weaknesses

1. What were the biggest benefits of working with GMI?
2. What challenges did you encounter in working with the program?

Suggestions for Improvement

1. In addition to your previous suggestions, do you have other suggestions for improvement?

Other Issues

1. Are there any other issues that we haven't touched upon that you would like to discuss?

Ideas for Establishing Performance Metrics for Gender Impacts

As discussed in Chapters Two and Four, OES/EGC requested information on GMI's contributions toward the following performance indicators:

- amount of greenhouse gas emissions, measured in metric tons of CO_2 equivalent, reduced or sequestered as a result of USG assistance
- number of people receiving training in global climate change (by gender)
- number of laws, policies, agreements, or regulations addressing climate change that have been proposed, adopted, or implemented as a result of USG assistance—specifically, those directly benefiting women or other marginalized groups
- number of institutions with an improved capacity to address climate change issues as a result of USG assistance—specifically, those serving women or other marginalized groups.

In this appendix, we identify potential methods for measuring the impacts that are specifically related to gender. The approaches that can be used to measure these effects will depend on the scope of the definitions chosen for each metric. Here, we describe a range of options, some of which are not mutually exclusive.

A basic measure of the *number of women trained in global climate change* could be the number of female beneficiaries (1) attending funded training programs or (2) exposed to funded demonstration projects or projects provided by organizations receiving indirect GMI support. An expanded definition could also include (3) the number of female employees attending funded staff training programs in global climate change; (4) the number of female beneficiaries/employees receiving any other form of funded training; (5) if general operating costs are supported by GMI, the number of female employees attending staff training programs in global climate change; (6) if partnerships are supported by GMI, number of female employees receiving training via partnerships or collaborations; and/or (7) the number of female beneficiaries/employees receiving any other form of training enabled by organizational support from GMI.

These measures may be obtained through a desk review of program documents and reports from each organization, either by external evaluators or via a written data requests to the organizations. If such records are not available, (self-reported) estimates may be collected via key informant interviews with officials at each organization. In each case, where possible, such estimates should be reviewed or validated by a knowledgeable external observer. A tighter focus on results might restrict the definition above to the number of female beneficiaries/employees in each of the categories who have demonstrated gains in skills, knowledge, or awareness. The most rigorous methodology requires the collection and analysis of existing pre-post administrative or survey data, comparing changes in knowledge among program beneficiaries to the changes of knowledge in a suitable comparison group. With no pre-program data, a feasible alternative may be the collection of new survey data, either comparing participants' levels of knowledge to that of a comparison group that was similar prior to the program or eliciting subjective perceptions of knowledge change. In circumstances in which a rigorous quantitative approach is not possible, rather than expend resources pursuing it, a more limited but appropriate qualitative approach may be to elicit such perceptions from focus groups. The *number of laws, policies, agreements or regulations addressing climate change that directly affect women that have been proposed, adopted, or implemented as a result of USG assistance* is possibly the most subjective and difficult to capture using a predefined methodology. In practical terms, this may be measured through a combination of individual consultations with funded organizations and discussions with experts. Recipient organizations can be asked to report and describe any laws, policies, agreements, or regulations that have been proposed, adopted, or implemented in the previous fiscal year, either implicitly or explicitly through their funded programs or indirectly funded efforts. Organizations would also be asked to justify this through channels such as published reports and policy briefs; participation in conferences, workshops, and meetings with policymakers; outreach efforts; and other campaigns. Local country experts could be asked to list important changes in the policy environment in the previous fiscal year related to climate change and to identify any programs or organizations that were associated with these changes.

Classifying such laws, policies, agreements and regulations as "directly affecting women" may include (1) having a specific mandate addressing women or gender equality, (2) being relevant to an industry or sector in which women are overrepresented, or (3) being relevant to gender-equalizing practices in an industry or sector in which women are underrepresented. This classification may be provided to respondents *ex ante* or may be applied *ex post*. An important trade-off is the degree to which an *ex ante* screening by respondents may be less labor-intensive for the evaluator, but it could result in errors of exclusion or inconsistent application.

It is important to recognize that this outcome measure is likely to be subject to errors of both inclusion and exclusion and that more precision in this case may not be achievable. In this instance, an evaluation may be practically restricted to the lesser objective of highlighting exceptional cases or examples.

Finally, in practice, the count of *organizations serving women with increased capacity to address climate change* may be most broadly interpreted, as it includes all funded organizations with a stated organizational mission of serving women or promoting gender equality, as well as those that have programs with such objectives. A desk review of program documents and reports from each organization could focus on these objectives. A more rigorous definition of serving women might restrict it to organizations in which the documented number of female beneficiaries exceeds a threshold number or percentage. A more rigorous definition of capacity-building might restrict it to organizations receiving direct funding for capacity-building activities or organizations receiving indirect funding that undertook capacity-building activities during the same fiscal year, including (1) staff training, (2) equipment purchases, and (3) partnership and collaboration. As with the previous metric, these data may be collected through a desk review or interviews with key staff, while a more outcome-driven definition geared toward measuring actual increases in capacity would require primary data collection, including systematic surveys of organization staff and equipment audits.

Site Visits: Site Selection Process, Protocol Development, and Contacting Respondents

Chapter Three provided some information on the country selection for our site visits, but we provide additional information in this appendix. We also describe how we developed the interview protocol that we used during the site visits and our process for contacting those we hoped to interview.

Site Selection

The selection process included the following steps:

1. We reviewed sector and emission profiles for all countries in which GMI has conducted activities and created a short list of countries that provided diversity in terms of sectors, geography, and emission levels. We then reviewed the GMI database to verify that short-listed countries had a sufficient volume of GMI activities.
2. We presented the short list of countries to EPA and requested feedback based on our selection criteria to ensure that we were not ignoring important details about GMI implementation in any one country that would affect project selection.
3. Based on feedback from EPA, we selected the final set of three countries (India, Mexico, and the Philippines).
4. We then reviewed the GMI activities conducted in each of the three countries and drew up a list of specific activities that were consistent with our overall criteria and specific goals for each country, which typically related to a diversity of activities within the sectors represented by that country.
5. We presented our initial list to EPA and asked for EPA's comments. Since we wanted to be as representative as possible given the small sample size, we were concerned only with activities that were deemed successful. We asked that EPA take into consideration activities that might be more fully representative when providing comments.
6. For each country we compiled a final list of GMI activities; we worked with EPA to contact the individuals associated with each activity in the three countries to set up meetings.

Protocol Development

During our site visits, we conducted semistructured interviews with key respondents: people who received funding for GMI activities; those who attended a training session, workshop, or meeting; local sector and government representatives that partner with GMI; and others. Given the variation across countries, GMI sectors, and the background of respondents, we were unable to generate a written survey that *a priori* would be applicable to all respondents in all situations. Therefore, we developed semistructured interviews that allowed for a two-way conversation between the interviewer and the interviewee. These conversations not only provided qualitative data on respondents' opinions, but they also helped the researcher understand the reasoning behind the respondents' statements. We specifically used probes and open-ended questions to explore reasoning. This form of data collection provided respondents with the flexibility to talk about their views in their own words. Doing so established a level of comfort that would have been difficult to establish using a more formal survey instrument, because we did not have prior connections to these respondents.

We constructed the protocol using a matrix format. We first identified the topical areas that were relevant to this evaluation and that we wanted to cover with the interviewees: project implementation, impacts, funding, other support received from GMI, and overall opinions about the program. For each of these areas, we were interested in understanding the what, when, how, and why. Since we also wanted to validate data from the GMI database, we posed questions to participants based on the nature of their involvement with GMI. For instance, if a participant had received GMI funding, we wanted to know whether he or she had received other sources of funding, why the participant applied for the funding, and the participant's experience with the funding process. The interview protocol can be found in Appendix B.

Process for Contacting Respondents

After selecting countries and sectors for site visits, we obtained contact information for key respondents from EPA officials. For all three site visits, EPA officials emailed respondents introducing the RAND team and notifying them that we would soon be in touch. After EPA had emailed the notifications, we contacted respondents either by telephone or email. In these conversations, we briefly explained the nature of our visits and some of the topics that we hoped to explore with them, and we arranged dates and times to meet in person.

Bibliography

Bäthge, Sandra, *Climate Change and Gender: Economic Empowerment of Women Through Climate Mitigation and Adaptation?* working paper, Eschborn, Germany: Deutsche Gesellschaft für Technische Zusammenarbeit, October 2010.

DoS—*See* U.S. Department of State.

EPA—*See* U.S. Environmental Protection Agency.

Forster, Piers, and Venkatachalam Ramaswamy, et al., "Changes in Atmospheric Constituents and in Radiative Forcing," in Susan Solomon, Dahe Qin, Martin Manning, Melinda Marquis, Kristen Averyt, Melinda M. B. Tignor, Henry LeRoy Miller, Jr., and Zhenlin Chen, *Contribution of Working Group I to the Fourth Assessment Report of the Intergovernmental Panel on Climate Change*, Cambridge, UK: Cambridge University Press, 2007, pp. 129–234.

Global Methane Initiative, *The U.S. Government's Global Methane Initiative Accomplishments*, October 2011. As of June 19, 2012:
http://www.epa.gov/globalmethane/pdf/2011-accomplish-report/usg_report_2011_full.pdf

———, "About the Initiative," web page, undated(a). As of June 19, 2012:
http://www.globalmethane.org/about/index.aspx

———, "Project Network," web page, undated(b). As of June 19, 2012:
http://www.globalmethane.org/project-network/index.aspx

GMI—*See* Global Methane Initiative.

Greenfield, Victoria A., Valerie L. Williams, and Elisa Eiseman, *Using Logic Models for Strategic Planning and Evaluation: Application to the National Center for Injury Prevention and Control*, Santa Monica, Calif.: RAND Corporation, 2006. As of June 19, 2012:
http://www.rand.org/pubs/technical_reports/TR370.html

Houghton, J. T., L. G. M. Filho, B. A. Callander, N. Harris, A. Kattenberg, and K. Maskell, eds., *Climate Change 1995: The Science of Climate Change: Contribution of Working Group I to the Second Assessment Report of the Intergovernmental Panel on Climate Change*, Cambridge, United Kingdom, Cambridge University Press, 1996.

International Institute for Energy Conservation, Eastern Research Group, and PA Consulting Group, *Resource Assessment for Livestock and Agro-Industrial Wastes—Philippines*, Methane to Markets Partnership, July 2, 2009. As of June 19, 2012:
http://www.globalmethane.org/documents/ag_philippines_res_assessment.pdf

Intergovernmental Panel on Climate Change, *IPCC Second Assessment: Climate Change 1995*, 1996. As of June 19, 2012:
http://www.ipcc.ch/pdf/climate-changes-1995/ipcc-2nd-assessment/2nd-assessment-en.pdf

M2M—*See* Methane to Markets Partnership.

Methane International, "Mumbai Landfill Awarded Advanced Carbon Credits," *Methane International*, No. 15, November 2009. As of June 19, 2012:
http://www.globalmethane.org/news-events/mi15.aspx#five

———, "India: Methane Project Opportunities in the Making," *Methane International*, No. 16, January 2010. As of June 19, 2012:
http://www.globalmethane.org/news-events/mi16.aspx#four

Methane to Markets Partnership, *U.S. Government Accomplishments in Support of the Methane to Markets Partnership*, October 2006. As of June 19, 2012:
http://www.epa.gov/globalmethane/pdf/2006-accomplish-report/M2M_accomplishments_full.pdf

———, *U.S. Government Accomplishments in Support of the Methane to Markets Partnership*, September 2007. As of June 19, 2012:
http://www.epa.gov/globalmethane/pdf/2007-accomplish-report/m2m_07_update_final.pdf

———, *The U.S. Government's Methane to Markets Partnership Accomplishments*, November 2008. As of June 19, 2012:
http://www.epa.gov/globalmethane/pdf/2008-accomplish-report/m2m08_usg_report_08_scrnrez.pdf

———, *Partnership Accomplishments, 2004–2009*, 2009a. As of July 19, 2012:
http://www.globalmethane.org/documents/par_110609.pdf

———, *The U.S. Government's Methane to Markets Partnership Accomplishments*, October 2009b. As of June 19, 2012:
http://www.epa.gov/globalmethane/pdf/2009-accomplish-report/m2m_usg_fullreport.pdf

———, *The U.S. Government's Methane to Markets Partnership Accomplishments*, November 2010. As of June 19, 2012:
http://www.epa.gov/globalmethane/pdf/2010-accomplish-report/usg_fullreport_2010.pdf

Reilly, John M., Henry D. Jacoby, and Ronald G. Prynn, *Multi-Gas Contributors to Global Climate Change: Climate Impacts and Mitigation Costs of Non-CO₂ Gases*, Arlington, Va.: Pew Center on Global Climate Change, February 2003.

UNDP—*See* United Nations Development Programme.

United Nations Development Programme, *Resource Guide On Gender and Climate Change*, 2009. As of June 19, 2012:
http://www.uneca.org/acpc/about_acpc/docs/UNDP-GENDER-CLIMATE-CHANGE-RESOURCE-GUIDE.pdf

United Nations Framework Convention on Climate Change, "Clean Development Mechanism (CDM)," web page, undated. As of June 19, 2012:
http://unfccc.int/kyoto_protocol/mechanisms/clean_development_mechanism/items/2718.php

U.S. Department of State, *U.S. Climate Action Report 2010: Fifth National Communication of the United States of American Under the United Nations Framework Convention on Climate Change*, Washington D.C.: Global Publishing Services, June 2010.

———, "Outcome Evaluation of the Methane to Markets Partnership Relative to the Office of Global Change Funding, FY2006–FY2010," request for applications, No. OES-OCC-11-004, July 11, 2011. As of June 19, 2012:
http://www.grants.gov/search/search.do?mode=VIEW&oppId=104453

———, "The Climate and Clean Air Coalition to Reduce Short-Lived Climate Pollutants," fact sheet, February 16, 2012. As of June 19, 2012:
http://www.state.gov/r/pa/prs/ps/2012/02/184055.htm

U.S. Environmental Protection Agency, *Global Anthropocentric Non-CO₂ Greenhouse Gas Emissions, 1990–2020*, Washington, D.C., 2006.

———, "Methodology for Estimating Leveraged Funding for USG Activities in Support of the Global Methane Initiative," November 2011a. Not available to the general public.

———, "US Government Efforts in Support of the Global Methane Initiative: Programmatic Metrics for Success," November 2011b. Not available to the general public.